ECOLOGY 2

FEBRUARY 2ND 1983

THE ECOLOGY OF
SCOTLAND'S LARGEST LOCHS
Lomond, Awe, Ness, Morar and Shiel

MONOGRAPHIAE BIOLOGICAE

VOLUME 44

Editor

J. ILLIES

Dr W. Junk Publishers, The Hague-Boston-London 1981

THE ECOLOGY OF SCOTLAND'S LARGEST LOCHS

Lomond, Awe, Ness, Morar and Shiel

Edited by
PETER S. MAITLAND

Dr W. Junk Publishers, The Hague-Boston-London 1981

Distributors:

for the United States and Canada

Kluwer Boston, Inc.
190 Old Derby Street
Hingham, MA 02043
USA

for all other countries

Kluwer Academic Publishers Group
Distribution Center
P.O. Box 322
3300 AH Dordrecht
The Netherlands

This volume is listed in the Library of Congress Cataloging in Publication Data

ISBN 90-6193-097-9 (this volume)
ISBN 90-6193-881-3 (series)

Dedicated to the staff of the Bathymetrical
Survey of the Fresh Water Lochs of Scotland
(1897–1909): a major initiative in limnology.

A satellite photograph of part of Scotland (scale 1 : 1 500 000), showing many of its important lochs and including Lochs Lomond, Awe, Ness, Morar and Shiel. For reference, see Fig. 1.2. (Photo: Royal Aircraft Establishment, Farnborough.)

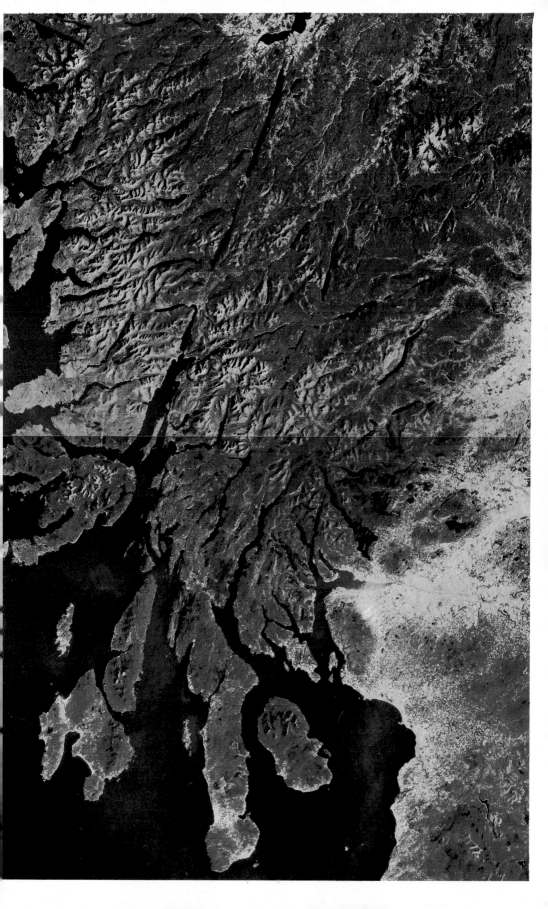

Preface

Scotland is fortunate in being very richly endowed with natural fresh waters in the form of numerous lochs and rivers. These constitute on the one hand an attractive feature of the landscape and on the other a major resource for industry and recreation. Thus there are about 3800 lochs over 4 ha in area and these form approximately 1.0% of the total surface area of Scotland. Comparable figures for England and Wales are 1700 lakes and 0.05% of the land surface, and in terms of volume, Loch Ness contains more water than all the lakes and reservoirs in England and Wales put together (Smith & Lyle 1979).

Many of the Scottish lochs are large and clean and consequently are particularly valuable in resource terms. The decision as to which are actually the largest lochs is debatable, particularly when the main criteria of greatest surface area, length, volume and depth each gives a different water (Lochs Lomond, Awe, Ness and Morar respectively). These four then are certainly among the largest lochs in the country, but close to them in size come several other large waters, among which is Loch Shiel – which is exceeded in length only by Lochs Awe, Ness and Lomond. These five very large lochs (Frontispiece) form the group of waters selected for the comparative studies described in this volume.

Although world-wide a number of very early freshwater studies are available (e.g. in Scotland: Anonymous 1755; Galbraith 1829) it was not until the late 19th century that limnology started to develop properly as a science. At around this time several classical studies were being innovated in both North America (Birge 1879) and Europe (Forel 1892). The bathymetrical survey of the freshwater lochs of Scotland was initiated by Sir John Murray in 1897 and widely recognised internationally (e.g. Collet 1908) at the time. It must be recognised as the start of organised limnology in the British Isles and many of the resulting publications (e.g. Murray & Pullar 1910) are still of major value.

Little comparative work on the large lochs of Scotland has been undertaken since these classical studies by Sir John Murray and his team – to whom this volume is dedicated. The present study partly originated in a request from the North of Scotland Hydro-Electric Board to carry out an impact assessment of a proposed pumped-storage hydro-electric station on the ecology of Loch Lomond. It was decided that one of the most useful ways of doing this would be to compare certain aspects of the ecology of this loch with those of the only two other lochs where pumped-storage schemes are already in existence in Scotland – Loch Awe and Loch Ness. At the same time as this project was being discussed a request came from the Nature Conservancy Council to carry out comparative ecological studies of Loch Morar and Loch Shiel – for conservation assessment purposes. It was decided then to embark on a broad comparative study of the ecology of all five waters – Scotland's largest lochs. The work is complimentary to another study (not reported here) dealing with broader aspects of the limnology of British fresh waters (Maitland 1979).

The research was carried out from 1977–80, mainly by staff from the Institute of Terrestrial Ecology. The project was partly funded by the North of Scotland Hydro-Electric Board (ITE Contract Project 546) and we are especially grateful to Mr P. L. Aitken, Dr J. Berry, Mr R. M. Jarvis, Mr S. D. Sedgwick and Mr M. W. C. Williams for their help and co-operation. Substantial support was also provided by the Nature Conservancy Council (ITE Contract Project 545) and help and advice was received from Dr J. M. Boyd, Mr R. N. Campbell, Dr D. A. Goode and Dr C. Newbold. Considerable assistance throughout the research and in the reviewing of manuscripts was also given by many other people who are acknowledged separately in the individual studies.

Much of the study, as is usual in such cases, was determined to a large extent by a number of logistic factors. The large surface areas of the lochs concerned, their great depths, the frequently arduous weather conditions and the driving distances from Edinburgh (a round trip of almost 1000 km is necessary to visit all the lochs) restricted the number of sampling trips and limited the field team to 4 persons. Thus, for example, the seasonal sampling frequency was less than ideal.

Some comment too must be made on the format of the present volume. It is presented as a series of ten independent but comparable ecological studies each of which includes an abstract and can be read on its own. Thus though the studies have been integrated as far as possible, cross-reference is minimal. This has led to some repetition of basic information concerning the lochs, but this was considered worthwhile in order to maintain the integrity of each study. It should be made clear that the main emphasis in the research was on land use, physical limnology, chemistry, phytoplankton, macrozooplankton, littoral invertebrates and fish. However, in order to make the comparisons more comprehensive, short review accounts of the macrophytes and profundal zoobenthos have also been included. No work was done on Bacteria, Protozoa and other important components of the loch communities. Nor was any attempt made to measure production or carry out many of the other interesting and important studies which require greater man-power or easier access to the lochs. The final synthesis is concerned with highlighting the main ecological features of the large lochs and comparing these with waters elsewhere. One of the points to emerge from the discussion in this section is that, in spite of the large bank of data now available, there is still much that is unknown about these large bodies of water.

References

Anonymous, 1755. Accounts of the earthquake (Lisbon) upon Loch Lomond (1755). Scots Mag. 17: 593–594.

Birge, E. A., 1879. Notes on Cladocera. Trans. Wisconsin Acad. Sci. Arts Lett. 4: 77–112.

Collet, L. W., 1908. Les Lacs d'Ecosse. Arch. Sci. Phys. Nat., Geneva. 25: 96.

Forel, F. A., 1892. Le Leman. Monographie limnologique. Lausanne. Vols. 1-3.

Galbraith, W., 1829. Barometric measurement of the height of Ben Lomond and on the quantity of water annually discharged by the River Leven from the basin of Loch Lomond. Edinb. New Phil. J. 6: 121.

Maitland, P. S., 1979. Synoptic limnology: the analysis of British freshwater ecosystems. Cambridge: Institute of Terrestrial Ecology. 28 pp.

Murray, J. & Pullar, L., 1910. Bathymetric survey of the freshwater lochs of Scotland. Edinburgh: Challenger. Vols. 1-6.

Smith, I. R. & Lyle, A. A., 1979. The extent and distribution of fresh waters in Great Britain. Cambridge: Institute of Terrestrial Ecology. 44 pp.

PETER S. MAITLAND

ZOOLOGY DEPARTMENT
UNIVERSITY OF ST, ANDREWS

Contents

List of contributors

Mrs Sheila M. Adair, Institute of Terrestrial Ecology, 78 Craighall Road, Edinburgh, Scotland.

Dr Antony E. Bailey-Watts, Institute of Terrestrial Ecology, 78 Craighall Road, Edinburgh, Scotland.

Mr Michael J. Carr, Springfield House, Oakfield Road, Gosforth, Newcastle-Upon-Tyne, England.

Dr Stephen P. Cuttle, Department of Forestry and Natural Resources, University of Edinburgh, Mayfield Road, Edinburgh, Scotland.

Ms Gillian M. Dennis, Department of Community Medicine, University of Edinburgh, Warrender Park Road, Edinburgh, Scotland.

Mrs Pamela Duncan, 10 Seaview Crescent, Edinburgh, Scotland.

Mr Alex A. Lyle, Institute of Terrestrial Ecology, 78 Craighall Road, Edinburgh, Scotland.

Dr Peter S. Maitland, Institute of Terrestrial Ecology, 78 Craighall Road, Edinburgh, Scotland.

Mr Andrew J. Rosie, Highland River Purification Board, Strathpeffer Road, Dingwall, Scotland.

Ms Barbara D. Smith, Institute of Terrestrial Ecology, 78 Craighall Road, Edinburgh, Scotland.

Mr Ian R. Smith, Institute of Terrestrial Ecology, 78 Craighall Road, Edinburgh, Scotland.

Dr Mark R. Young, Department of Zoology, The University, Aberdeen, Scotland.

1. Introduction and catchment analysis

P. S. Maitland

Abstract

This study is based on desk work which provides the background for a series of field studies carried out during 1977–80 on Scotland's largest lochs. Analyses of appropriate maps and tabulations of other relevant data are used here to assess the nature of these lochs and the contribution from and variety of their catchments. Human activities likely to affect the ecology of the lochs are considered in some detail. The results indicate a contrast between Lochs Lomond, Awe and Ness, naturally rich, likely to be further enriched by human activities in their catchments and subject to a variety of human pressures, and Lochs Morar and Shiel, naturally poor in nutrients with relatively little cultural eutrophication (especially Loch Morar) from the catchments and few human pressures on their systems. The relevance of these differences to the ecology of the lochs is discussed.

Introduction

The tremendous importance of Scotland's large freshwater lochs, though sometimes appreciated aesthetically, is often forgotten in terms of a natural resource. As well as their value to tourism these waters are a major asset to sport and commercial fisheries, to water-based recreation, to water supply, to the production of electricity (Aitken 1956) and in some cases to transport (Lindsay 1968). In times of critical water shortage in other parts of Great Britain, Scotland is fortunate in having enormous resources of high-quality water, for in relation to its size (34% of the land surface of Great Britain) it includes a very high number of both standing waters over 4 ha (3788) and running waters (3957) representing 69 and 51% respectively of the British

Monographiae Biologicae, Vol. 44, ed. by P. S. Maitland
1

totals (Smith & Lyle 1979). The largest river by flow (the River Tay) and all the largest lochs occur here: apart from the five lochs which form the subject of this study there are at least another 20 whose surface areas exceed 8 km².

In spite of their national importance, and the fact that they are being subjected to increasing human pressures (both to the waters themselves and via their catchments) relatively little attention has been paid to the ecology of the large lochs of Scotland. Because of the pressures on them it is important to review regularly their status and value to different sectors of the community. Their scientific merit is only too often given insufficient priority. The project introduced in this paper is the first comparative account of several of the largest lochs since the classical work of Murray & Pullar (1910).

The project arose because of the need for a comparative study of large oligotrophic lochs of Scotland, following the proposal by the North of Scotland Hydro-Electric Board to build a large pumped-storage power station at Craigroyston on Loch Lomond. In view of its international importance as a wetland site, and its national status as a Grade 1 open water and National Nature Reserve (Ratcliffe 1977) it was decided to study the possible impact of the Craigroyston scheme on Loch Lomond. Only two pumped-storage power stations exist in Scotland at present – at Cruachan on Loch Awe and at Foyers on Loch Ness – and it was felt that in view of the lack of information (Berry 1955) on the effect of such developments some research on the ecology of both these lochs was a major part of assessing the future of Loch Lomond with a similar hydro-electric scheme. At the same time as the study of these three large lochs was being considered, the need for basic ecological information on two others of importance (Lochs Morar and Shiel) arose. The value of a combined multidisciplinary study of all five lochs was at once apparent and the present study was initiated.

The work was carried out under contract to the North of Scotland Hydro-Electric Board and the Nature Conservancy Council and was limited mostly to a period of 18 months. Field work started in october 1977 and was mostly completed by October 1978 – though a few projects continued beyond this date. The basis of the comparative studies was a series of nine trips during this year around all the lochs – each trip allowing one day at each loch, usually in the following sequence: Ness Morar, Shiel, Awe and Lomond. Studies were carried out of littoral substrates, water temperatures, water chemistry, plankton and zoobenthos, and in addition all other available information (most of it unpublished) was brought together for analysis. A number of individual studies related to the effect of pumped-storage schemes was also carried out. These are not included in this series of papers.

This paper is the introductory one of a series describing these multidisci-

Plate 1.1 Vertical aerial photograph (scale: 1 : 11 000) of part of the Loch Lomond catchment, showing the main types of catchment land use mentioned in the text – open water, forest, urban development, rough and arable ground (Photo: Scottish Development Department.)

plinary studies of Scotland's largest lochs: Lomond, Awe, Ness, Morar and Shiel. Other papers in the group deal with comparative aspects of their physical limnology (Smith *et al.* 1981a), chemistry, phytoplankton and macrophytes (Bailey-Watts & Duncan 1981a, b, c), zooplankton (Maitland *et al.* 1981a), littoral zoobenthos (Smith *et al.* 1981b), profundal zoobenthos (Smith *et al.* 1981c), fish (Maitland *et al.* 1981b) and synthesis (Maitland *et al.* 1981d). Several other papers deal with related work (Maitland *et al.* 1981c, George & Jones 1981). The present account is basically a desk study of the five large lochs in relation to their catchments and local human activities: this was felt to be an essential background to the other comparative ecological studies being carried out.

Methods

The catchment analysis was based mainly on studies of the 1 : 50 000 Ordnance Survey maps of the areas concerned, supplemented for some features by the 1 : 63 360 O.S. maps and appropriate geological maps. Before map analysis started, the catchment of each loch was clearly drawn out on a master 1 : 50 000 O.S. map. In addition, each catchment was divided into four sub-catchments for comparative studies. These sub-catchments were very approximately equal in area for each loch: their actual boundaries were determined by hydrological, geological and land use features of potential importance to the loch.

Data were recorded from each one kilometre square of the catchment on to prepared sheets. All squares wholly within the catchment were analysed as well as those through which the watershed line passed which were more than half inside. The others were rejected. The catchments studied were the natural ones for each of the lochs concerned and did not take into account man-made alterations known to have occurred at Lochs Lomond, Awe and Ness.

The map parameters chosen for analysis and measured for each square were those felt to be most significant to the ecology of the lochs. They are as follows: (1) Altitude – the height in metres above sea level of the square centre. (2) Slope – the most typical slope in metres per kilometre within the square. (3) Catchment land use – the area in hectares occupied by open water, forest, urban development, rough and arable ground, and other types of land use (Plate 1.1). (4) Lochs – the number in each square. (5) Roads – the length in kilometres of metalled road in each square. (6) Aspect – the presence of significant areas of slope facing north, south, east or west. (7) Stream junc-

4

tions – the number in each square. (8) Waterfalls and wells – the number in each square. (9) Houses – the number marked in each square.

These final data were then summed to provide total or mean values for each sub-catchment, and for each catchment. The solid geology of each sub-catchment was measured by estimating from geological survey maps the percentage of base-rich rocks present. All conversions from sub-catchment to catchment values are weighted according to the proportionate area of each sub-catchment.

Human pressures on these large lochs were considered under three headings – those affecting the catchments, the lochs, or their outflows respectively. The first two of these are probably the most important as far as the ecology of the lochs is concerned, but pressures on the outflow are certainly relevant to migratory fish populations and some other features. The information for this part of the study was obtained mainly from the literature, from unpublished reports and direct from appropriate organisations. Thus the data on sewage and industrial discharges was taken from the recent report by the Scottish Development Department (1976), and on hydro-electric installations from the report by the North of Scotland Hydro-Electric Board (1978). Unpublished information was obtained from local River Purification Boards, the British Waterways Board (Loch Ness) and various other organisations, as well as the literature (Cormie 1970; Huxley 1979).

The large lochs

The definition of what is the largest freshwater loch in Scotland is debatable, particularly when the usual criteria of greatest surface area, length, volume and depth each gives a different water (Lochs Lomond, Awe, Ness and Morar respectively). A study of the data provided by Murray & Pullar (1910) shows that if the 12 largest lochs in each of the five categories: length, surface area, maximum depth, mean depth and volume are compared then 16 waters come into consideration (see Maitland 1976). These are listed in Table 1.1 with their relevant sizes. Their locations are shown on Fig. 1.1, from which it can be seen that all of them lie north of the Highland Boundary Fault Line and on the mainland of Scotland. Lochs Lomond, Awe, Ness and Morar are clearly the largest lochs in some respects. The fourth largest loch by length is Loch Shiel, and as this is one whose water regime is among the least affected by man it was included for comparative purposes in the group surveyed in this study. The major features of these five large lochs are shown in Table 1.2.

Loch Lomond (Plate 1.2a) is probably the best known Scottish loch and is

Table 1.1 The largest Scottish lochs and their dimensions. The figures are adapted from Murray & Pullar (1910) and do not include recent changes (e.g. at Loch Shin)

Loch	Length (km)	Area (km^2)	Max depth (m)	Mean depth (m)	Volume (m^3.10^6)
1. Lomond	36	71	190	37	2628
2. Awe	41	39	94	32	1230
3. Ness	39	57	230	132	7452
4. Morar	19	27	310	87	2307
5. Shiel	28	20	128	41	793
6. Lochy	16	15	162	70	1132
7. Treig	8	6	133	63	417
8. Tay	23	26	155	61	1697
9. Shin	28	23	49	16	371
10. Ericht	23	19	156	58	1141
11. Katrine	13	12	151	61	818
12. Maree	22	29	112	38	1156
13. Rannoch	16	19	134	51	1032
14. Arkaig	19	16	110	47	797
15. Glass	7	5	111	49	248
16. Earn	10	10	88	42	433

Table 1.2 Details of some of the main features of the large lochs discussed in this study

Feature	Lomond	Awe	Ness	Morar	Shiel
National grid reference	263 598	270 017	285 429	177 690	178 072
Length (km)	36.4	41.0	39.0	18.8	28.0
Breadth (km) (mean)	1.95	0.94	1.45	1.42	0.70
Shoreline length (km)	153.5	129.0	86.0	60.0	85.2
Area (km^2)	71.1	38.5	56.4	26.7	19.6
Catchment area (km^2)	781	840	1775	168	248
Rainfall (cm/annum)	220	256	186	242	267
Island area (km^2)	4.67	0.32	0.01	0.27	0.11
Maximum depth (m)	189.9	93.6	229.8	310.0	128.0
Mean depth (m)	37.0	32.0	132.0	86.6	40.5
Altitude (m)	7.93	36.2	15.8	10.1	4.5
Volume (m^3.10^8)	26.28	12.30	74.52	23.07	7.93

the most southerly of the five considered here. Its main axis is north to south and has a maximum length of 36.4 km – only exceeded by Lochs Awe and Ness. The northern portion of the loch is long (about 22.3 km) and narrow (averaging just over one kilometre), but south of Ross Point it opens out and

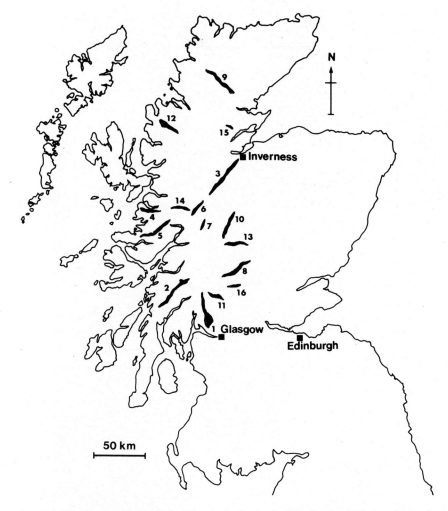

Fig. 1.1 The geographic position of the sixteen largest lochs in Scotland. The numbers refer to the lochs named in Table 1.1.

reaches a maximum width of 8.8 km. This part of the loch contains many islands. The surface area of Loch Lomond is 71.1 km² – larger than any other British standing water, the nearest rivals being Loch Ness (56.4 km²) and Loch Awe (38.5 km²). The catchment area of 781 km² is about ten times the surface area of the loch. The maximum depth is 189.9 m (the third deepest in Scotland) but the mean depth is only 37.0 m, because of the influence of the large shallow southern basin. The volume of Loch Lomond is 26.279 × 10⁸ m³, greater than any other lochs except Loch Ness. There are two distinct

7

Plate 1.2a Loch Lomond, viewed from near the north end of the loch looking north-east. (Photo: Scottish Tourist Board.)

8

Plate 1.2b Loch Awe, looking south-east across the northern end of the loch. (Photo: Scottish Tourist Board).

9

Plate 1.2c Loch Ness, looking west across the northern basin of the loch. (Photo: Scottish Tourist Board.)

Plate 1.2d Loch Morar, viewed from the north-west corner of the loch looking east. (Photo: Scottish Tourist Board.)

11

Plate 1.2e   Loch Shiel, looking south-west along the loch from the north end. (Photo: Scottish Tourist Board.)

12

basins deeper than 30 m in Loch Lomond, one to the south of the Douglas Water delta and one to the north and these are known to have quite distinct characteristics. The geology of the area is well known (Gregory 1928).

Loch Awe (Plate 1.2b) is the longest loch in Great Britain with a total length of 41.0 km. It is, however an extremely narrow loch, and is in most places only about one kilometre in breadth. The mean breadth is only 2.3 per cent of the length which is the smallest percentage observed in any of the 562 lochs surveyed by Murray & Pullar (1910). The surface area is 38.5 km² which means that Loch Awe is third largest in Great Britain in this respect. It has a mean depth of 32.0 m a maximum of 93.6 m and a total volume of 12.304×10^8 m³. Like Loch Lomond there are two main basins deeper than 30 m – one occupying the northern arm (with two separate depressions), the other occupying most of the long axis of the loch from the northern islands south.

Loch Ness (Plate 1.2c) has probably the strongest claim to be the largest body of water in Great Britain. Its total volume is 74.519×10^8 m³, which is three times as large as the next biggest by volume (Loch Lomond). Loch Ness also has the greatest mean depth (132.0 m) of the Scottish lochs and is second in respect of length (39.0 km), area (56.4 km²), and maximum depth (229.8 m). Its enormous basin is a very simple one, with, in most places, very steeply shelving sides. The gradient is commonly more than 1 in 1, and often precipitous. Thus it has a very small shore zone for its surface area – smaller, in fact, in proportion than any other measured Scottish loch. The deposition of material opposite the entrance to the River Foyers divides the loch into two basins deeper than 180 m, each with very similar characteristics. Loch Ness runs in a north-east/south-west direction and forms nearly one half of the waterway known as the Caledonian Canal. This canal is described in detail by Lindsay (1968).

Loch Morar (Plate 1.2d) is the deepest loch in the British isles with a maximum depth of 310.0 m. It is elongate in shape with its main axis running in an east-west direction. It has a total length of 18.8 km, a mean depth of 86.6 m and a volume of 23.073×10^8 m³. Though there are two basins both deeper than 270 m Loch Morar is essentially a single large basin, with, in most places very steeply shelving sides.

Loch Shiel (Plate 1.2e) has a total length of 28.0 km – only exceeded by Lochs Awe, Ness and Lomond. Like Loch Awe, it is very narrow with a mean breadth which is only 2.5% of its total length. The main upper portion of the loch runs in a north-east/south-west direction, but about 8 km from the outflow the axis bends and the lower portion bends almost due west. The surface area of the loch is 19.6 km²: it has a mean depth of 40.5 m and a maximum one of 128.0 m. It contains a total volume of 8.925×10^8 m³. Like

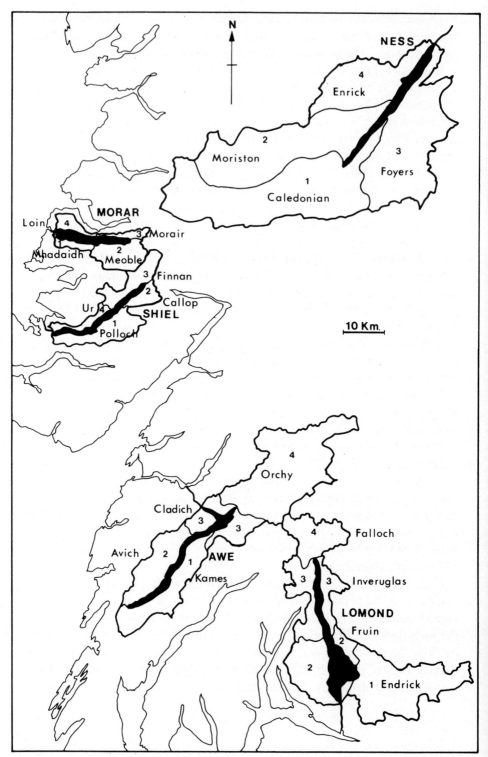

Fig. 1.2 The catchments of Lochs Lomond, Awe, Ness, Morar and Shiel, showing the sub-catchments analysed for this study.

Loch Lomond the northern end of Loch Shiel is much deeper than the southern end. The basin is essentially a single one, but there are a number of minor depressions here and there, two of them deeper than 120 m.

Sub-catchments

The divisions between the sub-catchments of the large lochs chosen for map study are shown in Fig. 1.2. The results of the analyses are given in Tables 1.3a-e. Though several of the sub-catchments show considerable similarity to each other, in some cases at least there are major differences in quality, and these are likely to be reflected in the hydrology and chemistry of the streams which drain them and subsequently in the loch basins into which they flow.

At Loch Lomond there is a major difference between the two northern sub-catchments (Falloch & Inveruglas) and the two southern ones (Fruin & Endrick) caused essentially by the Highland Boundary Fault Line. Those to

Table 1.3a   Results of the desk analysis of the four main sub-catchments of Loch Lomond

Feature	Lomond sub-catchments				Totals
	1. Endrick	2. Fruin	3. Inveruglas	4. Falloch	
Area (km^2)	264	161	158	113	696
Mean altitude (m)	172	190	311	438	249
Mean slope (m/km)	86	178	278	254	180
Land use (km^2)					
Rough	130.3	111.8	130.2	109.2	481.5
Water	1.3	0.3	3.4	0.6	5.6
Arable	92.3	21.4	2.1	0.2	116.0
Urban	5.5	1.2	4.0	0.1	10.8
Forest	32.1	23.8	18.1	2.7	76.7
Other	2.4	2.0	0.1	0.4	14.9
Rock (% base rich)	98	35	3	3	45
Aspect (% composition)					
North	28	21	23	21	24
South	26	27	23	29	26
East	21	27	28	27	25
West	25	25	26	23	25
Lochs	24	9	12	31	76
Stream junctions	204	266	224	148	842
Waterfalls	2	3	1	2	8
Houses	272	223	109	22	626
Metalled road (km)	1252	491	291	99	2133

Table 1.3b Results of the desk analysis of the four main sub-catchments of Loch Awe

Feature	Awe sub-catchments				Totals
	1. Kames	2. Avich	3. Cladich	4. Orchy	
Area (km^2)	148	118	116	398	780
Mean altitude (m)	234	199	245	392	307
Mean slope (m/km)	140	163	147	249	198
Land use (km^2)					
Rough	104.4	50.8	99.8	341.9	596.9
Water	2.7	5.2	2.2	4.4	14.5
Arable	4.2	1.8	3.8	2.7	12.5
Urban	0.5	0.4	1.2	0.8	2.9
Forest	36.2	59.4	8.6	48.3	152.5
Other	0.4	0.0	0.0	0.0	0.4
Rock (% base rich)	50	60	55	20	41
Aspect (% composition)					
North	30	22	26	26	26
South	20	27	24	24	24
East	20	28	23	26	24
West	30	23	27	24	26
Lochs	74	28	10	44	156
Stream junctions	196	157	95	801	1249
Waterfalls	0	0	2	3	5
Houses	102	87	122	96	407
Metalled road (km)	41	38	36	58	173

the north have high mean altitudes (both over 300 m), greater slopes (more than 250 m/km, considerable areas of rough ground and very little arable land. There is relatively little road and few houses. The southern sub-catchments, on the other hand, have much lower mean altitudes (both less than 200 m) more gentle slopes (less than 180 m/km), smaller relative amounts of rough ground and much more arable land. There is relatively more road, and many more houses. There is much more base rich rock in the southern sub-catchments than in the northern ones. In view of this and the main flow of water through the loch, one would expect the northern basin to be much less rich than the southern one.

At Loch Awe the main distinction in the sub-catchments lies between that of the River Orchy and the other three. The River Orchy catchment is very highland in character with a high mean altitude (392 m and slope 249 m/km) and a base-poor geology. The three other catchments in contrast, are much lower in altitude (all with means of less than 250 m) with gentler slopes (all less

16

Table 1.3c   Results of the desk analysis of the four main sub-catchments of Loch Ness

Feature	Ness sub-catchments				Totals
	1. Cale-donian	2. Moriston	3. Foyers	4. Enrick	
Area (km^2)	634	512	404	177	1727
Mean altitude (m)	357	375	473	296	384
Mean slope (m/km)	198	183	151	136	176
Land use (km^2)					
Rough	479.3	418.7	323.1	132.9	1354
Water	41.9	22.5	8.5	3.3	76.2
Arable	0.1	2.5	17.7	12.3	32.6
Urban	3.6	1.7	1.4	0.8	7.5
Forest	103.8	64.7	53.2	26.2	247.9
Other	5.3	2.3	0.5	1.9	10.0
Rock (% base rich)	5	8	30	15	13
Aspect (% composition)					
North	30	23	32	25	28
South	23	31	17	27	24
East	23	28	19	25	24
West	24	18	32	23	24
Lochs	67	108	93	87	355
Stream junctions	1047	542	342	116	2047
Waterfalls	28	10	5	1	44
Houses	131	245	359	299	1034
Metalled road (km)	82	106	116	64	368

than 170 m/km) and much richer geology. Thus it would be expected that the southern basin is richer than the northern one, though because of the position of the outflow considerable mixing must take place at some point in the northern basin.

At Loch Ness there is less contrast among the four sub-catchments. The greatest distinction lies between the two northerly (Enrick & Foyers) and the two southerly ones (Caledonian & Moriston) the former having relatively more arable ground and a richer geology, the latter having less arable land and a poorer geology. The northern basin of the loch could be expected to be richer than the southern one, though the main flow of water through the loch from south to north will tend to minimise this.

At both Loch Morar and Loch Shiel there are relatively few differences among the sub-catchments concerned, though there is a tendency for there to be more people and more arable ground nearer the western end (where the

Table 1.3d Results of the desk analysis of the four main sub-catchments of Loch Morar

Feature	Morar sub-catchments				Totals
	1. Mhadaidh	2. Meoble	3. Morair	4. an Loin	
Area (km²)	19	74	19	29	41
Mean altitude (m)	184	313	233	160	247
Mean slope (m/km)	203	376	377	163	303
Land use (km²)					
Rough	17.1	70.0	18.9	26.7	132.7
Water	0.2	1.3	0.1	1.2	2.8
Arable	0.1	0.4	0.1	0.3	0.9
Urban	0.1	0.2	0.1	0.2	0.6
Forest	1.1	2.4	0.1	0.9	4.5
Other	0.8	0.0	0.0	0.0	0.8
Rock (% base rich)	0	0	0	0	0
Aspect (% composition)					
North	32	32	13	15	25
South	19	26	54	39	32
East	28	19	20	17	20
West	21	23	13	29	23
Lochs	19	29	3	22	73
Stream junctions	14	111	29	28	182
Waterfalls	1	1	5	1	8
Houses	7	17	5	25	54
Metalled road (km)	0	0	0	3	3

outflows occur). There is thus likely to be less chemical variation among their waters than in the other three large lochs.

Fig. 1.3 shows a triangular plot of the sub-catchment data for the percentage composition by the three main types of land use: rough (plus water), forest (plus other) and arable (plus urban). It can be seen that most of the sub-catchments are grouped towards the nutrient poor corner of the triangle. The two most unusual sub-catchments are Lomond 1 (Endrick) and Awe 2 (Avich) which have exceptionally high percentages of arable and afforested ground respectively.

Catchments

The mean data from the analysis of the catchments of all five lochs are presented in Table 1.4. Apart from differences in absolute catchment size,

18

Table 1.3e Results of the desk analysis of the four main sub-catchments of Loch Shiel

Feature	Shiel sub-catchments				Totals
	1. Polloch	2. Callop	3. Finnan	4. Ur	
Area (km²)	95	38	72	30	235
Mean altitude (m)	236	290	336	150	263
Mean slope (m/km)	340	409	485	236	380
Land use (km²)					
Rough	64.8	30.7	67.9	22.6	186.0
Water	2.4	0.5	0.3	0.2	3.4
Arable	0.5	0.1	0.0	2.7	3.3
Urban	0.5	0.2	0.2	0.4	1.3
Forest	23.7	6.0	3.6	3.6	36.9
Other	2.9	0.0	0.1	0.0	3.0
Rock (% base rich)	0	0	0	0	0
Aspect (% composition)					
North	30	29	24	22	27
South	21	20	27	27	23
East	24	24	25	27	25
West	25	27	24	24	25
Lochs	34	4	11	11	60
Stream junctions	228	82	138	44	492
Waterfalls	1	0	0	0	1
Houses	57	5	24	32	118
Metalled road (km)	10	4	4	9	27

there are also major distinctions among them as far as topography and land use are concerned. These are likely to be reflected by marked differences in the quality of the waters of the large lochs themselves.

Loch Lomond's catchment is characterised by a relatively low mean altitude and gentler slopes with a high percentage of arable ground and base-rich rocks. There are far more roads than in any of the other catchments and a relatively high population. The catchment of Loch Awe has a relatively high mean altitude and moderate slopes. Almost 20% of its land is afforested and much of its geology base-rich. Loch Ness's catchment has the highest mean altitude of all the lochs but the gentlest slope. There is an unusually high number of waterfalls on streams in its catchment. Loch Morar and Loch Shiel have very similar catchments in many respects. Both have moderate altitudes, but steep slopes, very little arable ground and base-poor geology. The main distinction between them is that there is much less forest around Loch Morar than Loch Shiel.

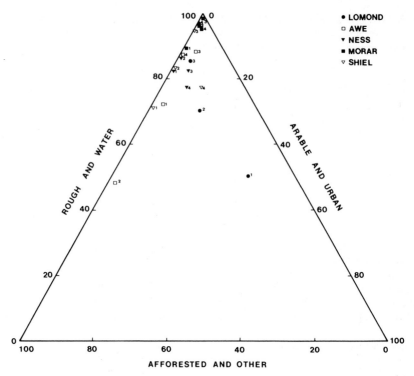

Fig. 1.3 A triangular diagram showing the land-use relationships among the sub-catchments of Lochs Lomond, Awe, Ness, Morar and Shiel.

Human pressures

Basic information on human pressures in the catchments, the lochs and their outflows is given in Tables 1.5, 1.6 and 1.7 respectively. The quality of the information in these tables varies but all of it is the most accurate and up-to-date available at the time of writing. Detailed interpretation is difficult because of the variation among the systems in catchment and loch size, rainfall, seasonal activity of tourists, boats etc., and probably only broad generalisations are of value at this stage.

The relevance of all the factors presented in these tables is worth commenting on. They fall into five basic categories though some of the factors are relevant to more than one of these. (1) Pollution, which must presumably be related to population size, sewage and industrial discharges, roads and boat activity. (2) Eutrophication, in addition to being linked with population size, sewage and some industrial discharges may also be considerably affected by

20

Table 1.4 Overall results from desk analyses of the large lochs catchments

Feature	Lomond	Awe	Ness	Morar	Shiel
Area (km²)	781	840	1775	168	248
Mean altitude (m)	249	307	384	247	263
Mean slope (m/km)	180	198	176	303	380
Land use					
(% composition)					
Rough	69.0	76.5	78.3	93.2	79.6
Water	0.8	1.8	4.4	2.0	1.5
Arable	17.0	1.6	1.9	0.6	1.4
Urban	1.5	0.4	0.4	0.4	0.6
Forest	11.0	19.6	14.4	3.2	15.7
Other	0.7	0.1	0.6	0.6	1.2
Rock (% base rich)	45	41	13	0	0
Aspect (% composition)					
North	24	26	28	26	27
South	26	24	24	32	23
East	25	24	24	21	25
West	25	26	24	21	25
Lochs	76	156	355	73	60
Stream junctions					
(nos/km²)	1.2	1.6	1.2	1.3	2.1
Waterfalls	8	5	44	8	1
Houses	626	371	1034	54	118
Metalled road (km/km²)	3.0	0.22	0.21	0.02	0.12

Table 1.5 Human pressures (and related data) on the catchments of the large lochs

Feature	Lomond	Awe	Ness	Morar	Shiel
Human population	12 218	777	3128	162	354
Sewage works dis-					
charges	7	2	8	0	0
Industrial discharges	2	0	0	0	0
Roads (km/km²)	0.30	0.22	0.21	0.02	0.12
Percentage arable					
ground	17.0	1.6	1.9	0.6	1.4
Percentage forest	11.0	19.6	14.4	3.2	15.7
Significant water					
abstractions	4	0	0	0	0
Significant water					
additions	1	2	1	0	0
Hydro-electric stations	1	1	5	0	0
Reservoirs	13	3	7	0	0
Waterfalls	8	5	44	8	1

Table 1.6 Human pressures (and related data) on the large lochs (direct)

Feature	Lomond	Awe	Ness	Morar	Shiel
Power boats: large	+	−	++	−	−
Power boats: small	+++	+	++	+	+
Non-power boats	+++	+	++	+	+
Sewage works discharges	1	0	2	0	0
Industrial discharges	0	0	0	0	0
Roads within 1 km (km)	54	83	66	5	12
Loch level control	+	+	+	+	−
Significant water abstractions	1	0	0	0	0
Significant water additions	0	0	0	0	0
Pumped-storage stations	0	1	1	0	0
Anglers (approximate)	1000	400	500	100	100
Angling season	11/2–31/10	11/2–15/10	2/2–15/10	11/2–31/10	11/2–30/10
Net fisheries	+	+	−	−	−

Table 1.7 Human pressures (and related data) on the outflows from the large lochs

Feature	Lomond	Awe	Ness	Morar	Shiel
Population within 2 km	40 000	60	35 000	70	100
Outflow length (km)	11	6	12	1	4
Sewage works discharges	2	0	1	0	0
Industrial discharges	3	0	1	0	0
Bridges	8	2	5	1	2
Obstructions	1	1	1	1	0
Hydro-electric stations	0	1	0	1	0
Anglers	+++	++	+++	+	++
Net fishery	—	—	—	—	—

the fertilisation of arable and afforested ground in the catchments. (3) Water regimes are also affected to some extent by catchment land use and reservoirs there but particularly by direct manipulation of water through abstraction, addition, and the activity of conventional and pumped-storage hydro-electric power stations. (4) Disturbance to the lochs by direct human activity is measured in part by the number of boats of different types and the extent to

Plate 1.3   Vertical aerial photographs (scale: 1 : 26 000) of the outflows from (a) Loch Lomond, (b) Loch Shiel. These show the enormous differences in human pressures on the outflows from these two lochs. (Photos: Scottish Development Department.)

23

which nearby roads make the lochs accessible to people (Tivy 1980). (5) Fish populations may be affected by the number of anglers and the extent of stocking, the presence of net fisheries, the length of the fishing season, and the number of obstructions (McGrath 1960) and bridges on the outflows (Plate 1.3).

Discussion

The desk analyses presented above indicate that there are a number of important differences among these lochs in spite of the fact that they are all large, long and deep. This introduction forms an essential background to the field studies described in other papers in the series and is important in assessing the significance and magnitude of the impact of any future developments – e.g. the proposed pumped-storage hydro-electric scheme at Craigroyston on Loch Lomond (Dill *et al.* 1971).

The analysis of sub-catchments shows that the greatest variety is found at Loch Lomond and the least at Loch Morar. Lochs Awe and Ness come second to Lomond in variation among sub-catchments, while Shiel is close to Morar in most features other than afforestation. The implication of the extent of the variety among the sub-catchment is that, not only may the stream types and communities within them vary, but the waters draining them may be responsible for variations in the lochs themselves. Thus much greater differences could be expected between the water and plankton communities of the north and south ends of Loch Lomond than the east and west ends of Loch Morar.

A comparison of the total catchments of the lochs shows that Loch Ness drains more than twice the area of the next largest catchment (Loch Awe). Loch Morar, with the smallest catchment (less than one tenth of that of Loch Ness), and Loch Shiel both drain relatively small areas. The effect of these differences on the hydrology of the systems concerned is discussed by Smith *et al.* (1981a). The extent of arbable ground and base-rich rocks gives a good indication of the potentital natural richness of the loch waters draining from them. It is clear from Table 1.4 that Loch Lomond (or certainly its south basin) seems likely to be the richest, followed by Lochs Awe and Ness. Loch Morar, and to a slightly lesser extent Loch Shiel, are likely to be very nutrient poor (see Bailey-Watts & Duncan 1981a, b).

The data on pressures on the large lochs, their catchments and outflows have been assessed in terms of precedence and grouped in terms of impact relevance in Table 1.8. It is clear that Lochs Lomond, Awe and Ness have

Table 1.8 Sums of rankings of precedence values for pressures of different types. Lowest values mean highest pressures

Feature	Lomond	Awe	Ness	Morar	Shiel
Pollution:					
Catchment	5	10	8	16	14
Loch	8	10	6	14	13
Outflow	3	11	6	10	9
Eutrophication					
Catchment	8	10	8	19	14
Loch	6	12	6	12	12
Outflow	2	8	4	7	7
Disturbance					
Catchment	2	5	5	10	8
Loch	8	13	9	18	17
Outflow	3	10	5	12	9
Hydrology					
Catchment	6	8	7	12	12
Loch	4	4	4	5	6
Outflow	3	2	3	2	4
Fisheries					
Catchment	8	12	8	13	15
Loch	7	11	8	12	12
Outflow	6	10	8	11	12
Totals	81	136	95	173	163

been subject to most human interference and loch Morar and Shiel (particularly the former) to least. The effect of this pressure can only be assessed by the appropriate ecological field studies, but one factor of major relevance is that the waters of Lochs Lomond, Awe and Ness are likely to be significantly enriched by the activities of humans in their catchment (Hasler 1947). Afforestation is the main factor affecting Loch Shiel, while Loch Morar is the most free of any potentital cultural eutrophication.

Acknowledgements

I am grateful to Ms. B. D. Smith, Mrs P. Duncan, Mr A. J. Rosie, Ms G. M. Dennis and Mr M. J. Carr for their help with much of the abstraction of catchment data. The drawings were prepared by Mrs S. M. Adair and the manuscript typed by Mrs M. Wilson.

References

Aitken, P. L., 1956. Hydro-electric schemes in the north of Scotland. I.U.C.N., Proc. Tech. Meet., Edinburgh, pp. 1–6.

Bailey-Watts, A. E. & Duncan, P., 1981a. The ecology of Scotland's largest lochs: Lomond, Awe, Ness, Morar and Shiel. Ed. P. S. Maitland. Chemical characterisation. A one year comparative study. Chap. 3. This volume.

Bailey-Watts, A. E. & Duncan, P., 1981b. The ecology of Scotland's largest lochs: Lomond, Awe, Ness, Morar and Shiel. Ed. P. S. Maitland. The phytoplankton. Chap. 4. This volume.

Bailey-Watts, A. E. & Duncan, P., 1981c. The ecology of Scotland's largest lochs: Lomond, Awe, Ness, Morar and Shiel. Ed. P. S. Maitland. A review of macrophyte studies. Chap. 5. This volume.

Berry, J., 1955. Hydro-electric development and nature conservation in Scotland. Proc. R. phil. Soc. Glasg. 77: 23–36.

Cormie, W. M., 1970. The Loch Lomond water scheme. J. Inst. Wat. Eng. 24: 291–318.

Dill, W. A., Kelley, D. W. & Fraser, J. C., 1971. The effects of water- and land-use development on the aquatic environment and its resources, and solutions to some of the generated problems. FAO Fish. Circ. 129: 1–7.

George, D. G. & Jones, D. H., 1981. Spatial studies of the zooplankton of Scotland's largest lochs: Lomond, Awe, Ness, Morar and Shiel. In preparation.

Gregory, J. W., 1928. The geology of Loch Lomond. Trans. geol. Soc. Glasg. 18: 301–323.

Hasler, A. D., 1947. The eutrophication of lakes by domestic drainage. Ecology. 28: 283–395.

Huxley, T. (Ed.), 1979. Shore erosion around Loch Lomond. Perth: Countryside Commission for Scotland.

Lindsay, J., 1968. The canals of Scotland. Newton Abbot: David & Charles.

McGrath, C. J., 1960. Dams as barriers or deterrents to the migration of fish. Proc. Int. Conf. Prot. Nat., Athens 4: 81–92.

Maitland, P. S., 1976. Fish in the large freshwater lochs of Scotland. Scott. Wildl. 12: 13–17.

Maitland, P. S., Smith, B. D. & Adair, S. M., 1981b. The ecology of Scotland's largest lochs: Lomond, Awe, Ness, Morar and Shiel. Ed. P. S. Maitland. The Fish and Fisheries. Chap. 9. This volume.

Maitland, P. S., Smith, B. D. & Dennis, G. M., 1981a. The ecology of Scotland's largest lochs: Lomond, Awe, Ness, Morar and Shiel. Ed. P. S. Maitland. The crustacean zooplankton. Chap. 6. This volume.

Maitland, P. S., Smith, I. R., Bailey-Watts, A. E., George, D. G., Lyle, A. A., Smith, B. D., Duncan, P., Rosie, A. J., Dennis, G. M. & Carr, M. J., 1981c. Twenty-four hour studies of the effects of pumped-storage power stations on water and plankton in Loch Awe (Cruachan) and Loch Ness (Foyers), Scotland. In preparation.

Maitland, P. S., Smith, I. R., Bailey-Watts, A. E., Smith, B. D. & Lyle, A. A., 1981d. The ecology of Scotland's largest lochs: Lomond, Awe, Ness, Morar and Shiel. Ed. P. S. Maitland. Comparisons and synthesis. Chap. 10. This volume.

Murray, J. & Pullar, L., 1910. Bathymetrical survey of the freshwater lochs of Scotland. Edinburgh: Challenger. Vols. 1–6.

North of Scotland Hydro-Electric Board, 1978. Power from the glens. Edinburgh: NSHEB.

Ratcliffe, D. A. 1977. A nature conservation review. Cambridge University Press. Vols. 1–2.

Scottish Development Department, 1976. Towards cleaner water 1975. Edinburgh: H.M.S.O.

Smith, B. D., Cuttle, S. P. & Maitland, P. S., 1981c. The ecology of Scotland's largest lochs:

Lomond, Awe, Ness, Morar and Shiel. Ed. P. S. Maitland. The profundal zoobenthos. Chap. 8. This volume.

Smith, B. D., Maitland, P. S., Young, M. R. & Carr, M. J., 1981b. The ecology of Scotland's largest lochs: Lomond, Awe, Ness, Morar and Shiel. Ed. P. S. Maitland. The littoral zoobenthos. Chap. 7. This volume.

Smith, I. R. & Lyle, A. A., 1979. The extent and distribution of fresh waters in Great Britain. Cambridge: Institute of Terrestrial Ecology.

Smith, I. R., Lyle, A. A. & Rosie, A. J., 1981a. The ecology of Scotland's largest lochs: Lomond, Awe, Ness, Morar and Shiel. Ed. P. S. Maitland. Comparative physical limnology. Chap. 2. This volume.

Tivy, J., 1980. The effect of recreation on freshwater lochs and reservoirs in Scotland. Perth: Countryside Commission for Scotland.

2. Comparative physical limnology

I. R. Smith, A. A. Lyle & A. J. Rosie

Abstract

The paper examines the physical conditions resulting from the interaction of loch structure with the external variables of rain, sun and wind. After a brief review of the climate, each of these variables is considered in turn and it is concluded that the external factors are comparatively constant and that many of the differences between the five lochs can be directly attributed to differences in morphometry. What variability there is, however, tends to enhance biological productivity in some lochs rather than others so that biological differences may be greater than examination of the physical characteristics alone might suggest.

Introduction

Physical limnology is taken here to mean the investigation of the environmental conditions that result from the interaction between the loch structure and the primary external forces of rain, sun and wind. This interpretation is used to define the layout of this part of the study, i.e. the climate, morphometry (structure), hydrology (rain), radiation and temperature (sun) and hydraulic conditions (wind). The final section is a review of the overall characteristics of the five lochs.

Climate

The climate of the British Isles is predominantly influenced by their position on the eastern seaboard of the Atlantic Ocean, adjacent to the European

Monographiae Biologicae, Vol. 44, ed. by P. S. Maitland

continental land mass. Consequently the day-to-day weather in Britain, and the west of Scotland in particular, is extremely changeable, being generally dominated by depressions and associated fronts travelling across the country from the west. Less frequent is the occurrence of anticyclonic continental air.

Characteristically depressions produce rain and wind throughout the year with low air temperature ranges and cloudy skies which reduce sunshine and radiation energy. In contrast, anticylonic air is calm and dry with clear skies and more extreme daily temperatures.

A description of wind conditions within the area is made difficult by the absence of anemograph records which are mainly taken for coastal sites. However, from those available, an examination of wind speed and direction frequency (Shellard 1968) clearly shows that west coast stations are more windy and reflect the dominant westerly trend in direction. The implications here may be misleading, however, when applied to mainland valleys because of the considerable effect of topography on local climate. Birse & Robertson (1970) in their assessment of climatic conditions in Scotland gave low values of exposure to the immediate vicinity of all five lochs, in contrast to the much harsher values given to the surrounding areas in general.

The mild winters and cool summers experienced close to the lochs are caused by the damping of seasonal fluctuations in air temperature not only by the proximity of the Atlantic Ocean but also by the heat absorption and release of the large volumes of water in the lochs themselves. Significantly ice is limited to shallow areas, even in severe winters. Average annual air temperatures of between 8 and 9 °C can be expected at low altitude in the area with a tendency for higher temperatures towards the west where the influence of the warm Gulf Stream current is strongest (Meteorological Office 1976a). August is generally the warmest month with mean daily air temperatures in the order of 14 °C and January the coldest with a mean of approximately 3 °C. During anticyclonic conditions, extreme summer maximum values may exceed 30 °C and winter minimums approach –20 °C. Air temperature is closely related to altitude, however, and average temperatures experienced in the higher catchment areas will be substantially cooler than those quoted above.

The approximate range in possible day-lengths at latitude 57 °N is 17.5 to 6.5 h for mid June and December respectively, the annual daily average being, of course, 12 h. Because of dull atmospheric conditions, the actual number of hours of recorded bright sunshine is only a fraction of this figure. (This is also partly due to the minimum input requirements of the recording apparatus). Daily averages for the region range between <3 and 3.5 h, with summer and winter values of approximately 6 and <1 h respectively. (Meteorological Office 1976b). The highest and lowest recorded accumulated monthly sunshine totals at Fort Augustus (57°09′ N) during the period 1941–70 were 252.3 h (July) and 9.0 h (December). The total possible day-length hours for these months are 530 and 209 respectively. Incoming solar radiation energy is directly related to the number of sunshine hours, but generalised calculations of this type can be inaccurate in upland areas however, since energy absorbtion and reflection varies greatly with the slope and aspect of the land. The loss of solar energy to the lochs by landscape shading is considered later.

A combination of open exposure to westerly winds travelling over the expanses of the Atlantic ocean, and a mountainous terrain ensure that the west of Scotland is the wettest area in Britain. The rainfall map published by the Meteorological Office indicates that upland areas experiencing rainfall in excess of 3000 mm annually are common and nowhere of specific interest here receives less than 1000 mm with the exception of the north half of Loch Ness. The dryest months are usually May or June with 5 to 6% of the annual total, the wettest being October or December with 11 to 12%, commonly there is a lapse in rainfall during November. The annual average rainfall has been calculated for the loch catchments and is given in Table 2.3.

An average of between 15 and 20 days of snow each winter can be expected in the lower valleys with an annual maximum of up to 40 days (Meteorological Office 1971). The occurrence and particularly the duration of snow fall is mostly dependent on air temperature which, as already shown is strongly related to altitude. Consequently the higher catchment areas of the lochs experience considerably more snow with an average of around 60 days and a maximum approaching 90. Although this snow cover acts as a reservoir of water which when released in spring can have an effect on stream flow, maximum flows are associated with high rainfall rather than with snow melt.

Morphometry

The information on the morphometric structure of the five lochs is all based on the surveys done by Murray & Pullar (1910). The only exception to this is that the mean height above sea level has been corrected using up-to-date water level records from various sources. No attempt has been made to correct any of the data derived from Murray and Pullar's surveys to take account of any of the small differences in mean water level. No difference in the comparative characteristics are likely to result from this although there will be slight errors in the absolute figures.

The structure of the lochs is illustrated in Fig. 2.1 and the main features

Table 2.1 Morphometric data for the five lochs

Feature	Lomond	Awe	Ness	Morar	Shiel
Length (km)	36.44	40.99	38.99	18.80	28.00
Mean breadth (km)	1.95	0.94	1.45	1.42	0.70
Water surface area (km²)	71.10	38.46	56.41	26.68	19.58
Mean depth (m)	37.0	32.0	132.0	86.6	40.5
Maximum depth	189.9	93.6	229.8	310.0	128.0
Volume m³ × 10⁹	2.628	1.230	7.452	2.307	0.793
Mean height above sea level (m)	7.92[1])	36.22[2])	15.80	10.13	4.54[3])
Shoreline length (km) including islands	153.48	128.97	86.00	60.03	85.18
Shoreline length (km) excluding islands	102.69	113.93	85.26	51.45	77.60
Area of islands (ha)	466.8	31.80	0.31	26.90	11.00
Distance from the sea (km)	30	20	25	2.5	6.3

[1]) Figure refers to water level after the construction of the River Leven Barrage.
[2]) Figure refers to water level after the construction of the River Awe Barrage.
[3]) Mean level is an estimate based on the relationship between Morar and Shiel levels.

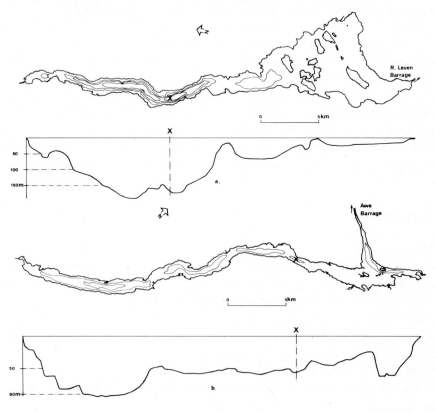

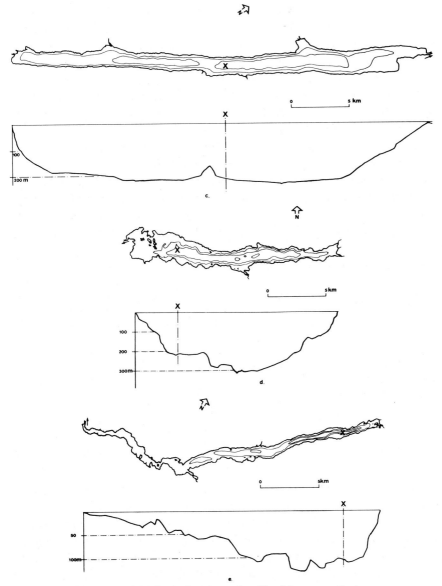

Fig. 2.1 Morphometry of the five lochs (plan and profile of deepest section).
(a) Loch Lomond, (b) Loch Awe, (c) Loch Ness, (d) Loch Morar and (e) Loch Shiel.
Standard sampling point is marked X on each map.

summarised in Table 2.1. The obvious feature, which is referred to throughout the entire account, is that, while four of the lochs can be treated as single

and reasonably homogeneous basins, Loch Lomond must be treated as at least two and, at times, three separate basins which have merged to form a single loch. Distance from the sea is defined as the shortest distance from the loch outlet to the open coasts, irrespective of direction. It is, inevitably, only an approximation because of the indented, fjord coastline. For all except Loch Ness, the open coast lies to the west or south west of the outlet, i.e. in the direction of the prevailing wind. Most of the other features are self-explanatory.

The shore zone, where the lochs and land impinge, is an important environment and its characteristics are variable and difficult to determine without detailed field survey. What has been done is to go back to the original Murray and Pullar surveys and measure the horizontal distance between the water's edge and the nearest sounding offshore, thus producing a measure of shore zone slope. The mean values for a number of transects are listed in Table 2.2.

Table 2.2 Shoreline characteristics for the five lochs

Loch	Number of transects	Mean shore zone slope m per m fall	Area of bed exposed per m fall ha m^{-1}
Lomond			
North Basin	22	9.91	45.97
Luss Basin	12	11.17	20.96
South Basin	30	28.78	255.26
Whole loch	64	18.99	291.86
Awe	51	14.09	182.18
Ness	39	3.71	33.04
Morar	26	9.92	59.85
Shiel	28	15.97	136.30

This is a considerable but necessary simplification as the degree of detail involved in separating the flatter slopes found near river deltas and the like from the 'hard edge' slopes elsewhere is not feasible in a comparative study. By combining these mean slopes with the shoreline length, it is possible to estimate the area of bed exposed for a given change in water level (Table 2.2). Again the features are more or less self-explanatory.

Hydrology

No attempt is made in this section to examine the hydrological behaviour of the lochs and their catchments in any detail. All that is considered are the broad features of the flow as they affect the chemistry of the lochs and the water level changes in the lochs themselves.

Streamflow

Loch inflows.   The division into sub-catchments of the areas draining into the five lochs has been described by Maitland (1981). The approximate mean flows from these sub-catchments have been estimated by subtracting an estimated constant evaporation (16 in. or 406.4 mm) from the estimated mean annual rainfall for the period 1916–1950 over the catchment. The results are listed in Table 2.3.

Table 2.3   Inflow Data for the sub-catchments of the five lochs

Loch	Catchment area km²	Mean annual rainfall mm	Estimated mean annual flow	
			m³ s⁻¹	% total
Lomond				
Endrick	266	1549	9.63	21.9
Fruin	196	2234	11.35	25.8
Inveruglas	175	2869	13.66	31.0
Falloch	113	3026	9.39	21.3
Awe				
Kames	149	2202	8.48	15.7
Avich	118	2314	7.13	13.2
Cladich	116	2414	7.38	13.7
Orchy	398	2855	30.89	57.3
Ness				
Caledonian	635	2176	35.63	42.3
Moriston	526	2285	31.33	37.2
Foyers	416	1368	12.68	15.1
Enrick	182	1185	4.49	5.3
Morar				
Mhadaidh	25	2139	1.37	12.9
Meoble	77	2688	5.57	52.3
Morair	25	2799	1.90	17.8
An Loin	36	1996	1.81	17.0
Shiel				
Pollach	101	2561	6.90	37.5
Callop	40	3175	3.51	19.1
Finnan	76	2858	5.91	32.1
Ur	35	2279	2.08	11.3

It must be emphasised that these are approximate figures but which are adequate for assessing the general features of the loch chemistry. In the case of Loch Lomond, for example, the estimated total flow is 44.03 m³s⁻¹. The estimated natural flow down the River Leven for the period 1951–1968, taken from the computer programme used to investigate the effects of the proposed Craigroyston Pumped-Storage Scheme on water levels, is 39.4 m³ s⁻¹, (North of Scotland Hydro-Electric Board, unpublished) i.e. a difference of 11.7%. This is partly because different time periods are involved but the evaporation may also be under-estimated, particularly for the Endrick sub-catchment. Since equivalent observed data are not available for the other lochs, manipulation of the evaporation figures is not justified and it is believed that the figure used is more correct for the more northerly catchments.

A feature of catchment hydrology which influences the chemistry of the lochs is the length of time the water stays on the land surface before being discharged to the stream system. The greater the delay, the greater the potential for chemical reactions and the greater the difference between the original rain water and the water entering the lochs. This is not always true since nitrogenous fertilisers can be washed out very rapidly and emphasises the importance of the proportion of arable land in the catchments.

Beran & Gustard (1977) define the Base Flow or Reliability Index as the proportion of the total flow that is derived from stored sources. An index value of 0.50 for example, implies that half the total flow has lain in the catchment for weeks or even longer and that half is flood water discharged within hours of falling as rain.

A few values of the index are available for catchments draining into the five lochs (Institute of Hydrology, unpublished). The general indication is that up to two thirds of the total inflow values are made up of recently discharged flood water and that this figure may rise even higher in some of the steep, hard rock sub-catchments.

Loading. Loch chemistry is affected not by the total volume of water discharged but by the solid and dissolved loads that the inflow water carries into the loch. Two figures quoted in Table 2.4 can be used to convert inflow chemical concentrations into absolute inputs.

The first of these figures is what is referred to as the unit loading, i.e. the quantity of solid or dissolved matter entering a loch if the mean concentration in the inflow stream is 1 mg l⁻¹. For example, if the mean inflow concentration is 0.1 mg l⁻¹, then the total quantity delivered per year into Loch Lomond is 138.95 tonnes. The second figure is the areal equivalent unit loading which is simply the previous figure divided by the surface area of the loch and ex-

Table 2.4 Chemical loading and retention times for the five lochs

Feature	Lomond	Awe	Ness	Morar	Shiel
Mean annual inflow m³ s⁻¹	44.03	53.88	84.13	10.65	18.40
Unit loading tonnes yr⁻¹	1389.5	1700.3	2654.9	336.1	580.7
Areal equivalent unit loading gm m⁻² yr⁻¹	19.5	44.2	47.1	12.6	29.7
Retention time yr	1.89	0.72	2.81	6.87	1.37

pressed as $gm\ m^{-2}\ yr^{-1}$. The theoretical retention time, i.e. the volume of the loch divided by the mean inflow rate, measures the theoretical mean length of time the water is retained in the loch. The actual length of time depends on mixing in the loch. In Loch Awe, for example, a high proportion of the total runoff enters at the north end from the River Orchy and some of this water may be discharged out of the loch without being completely mixed in the entire loch. It is shown later, however, that the volumes of water transported by wind-induced currents can be very much greater than the average flows. 'Short circuiting', i.e. the discharge from the loch of recently entered river water, is likely to be transient and dependent on appropriate combinations of high flows and low wind speeds.

An obvious feature of Table 2.4 is the almost ten-fold range in retention time. This is the most obvious distinguishing feature of the hydrological characteristics of the five lochs since the areal equivalent loading is less variable.

Loch water levels

Loch levels fluctuate because the inflow and outflow rates are not equal, the rise and fall being equal to the change in the volume of water stored in the loch. With a rising inflow, the usual effect of storage changes in the loch is to reduce the magnitude of the outflow compared to the inflow and to delay the occurrence of the peak. The main loch characteristics which determine the extent of these damping and lag effects are the loch area in relation to the total inflow and the form of the curve that relates the outflow rate to the loch level. The significance of the loch area is obvious since it alters the volume of water stored for the same rise in level. If the outflow rate is markedly increased for a small rise in level, then the level fluctuations will be damped whereas a constricted outlet will increase them. The influence of loch storage on falling flows is less pronounced.

Plate 2.1 Views of the outlets of the five lochs.

a. Barrage controlling the level and outflow of Loch Lomond for water supply purposes, looking towards the loch. (Photo: B. D. Smith.)

b. Barrage controlling the level and outflow of Loch Awe. Discharged water is used to generate hydroelectric power. A fish pass and counter are installed in the structure. (Photo: B. D. Smith).

c. Vertical aerial view of the Loch Ness outlet (scale: 1 : 26 165) showing part of the Caledonian Canal running alongside the River Ness. (Photo: Scottish Development Department.)

d. Control works at the outlet of Loch Morar. Discharged water is used to generate hydro-electric power. A Fish pass and counter are installed in the structure. (Photo: P. S. Maitland.)

e. The natural outlet channel of Loch Shiel, flow being from left to right. (Photo: K. H. Morris.)

39

Since the construction of the River Leven Barrage, only Loch Shiel has a natural outlet. Loch Awe is regulated by a barrage; Loch Ness is affected by the works associated with the Caledonian Canal and the short outlet channel from Loch Morar is influenced by a small hydro-electric scheme. Views of the outlets are shown in Plate 2.1.

Sources of data. None of the water level records has been published. Data for Loch Lomond are divided into two sets: the records from Balloch for the period 1955–1964 which relate to conditions before the construction of the water supply barrage on the River Leven; the records from Balmaha provided by the Nature Conservancy Council for the period 1973–1977 which relate to conditions when the flow down the River Leven is regulated by the barrage. There are no known water level records for completely natural conditions in the loch before the construction of the Loch Sloy Hydro-Electric scheme in 1948.

The records for Lochs Awe and Morar are based on observations by the North of Scotland Hydro-Electric Board while those for Loch Ness are based on the British Waterways Board gauge at Fort Augustus. No long-term

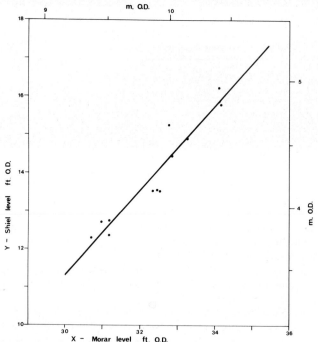

Fig. 2.2 Relation between the monthly mean levels in Lochs Morar and Shiel. $Y = 1.105X - 21.838$ (feet units)

40

records for Loch Shiel are known to exist but a gauge was installed at the head of the Loch in November 1977. This has been used primarily to establish relationships between levels in Lochs Morar and Shiel in order to generate a synthetic record for Loch Shiel.

The computerised data for Lochs Lomond, Ness and Morar have been analysed by the North of Scotland Hydro-Electric Board. The record for Loch Awe is in the form of charts and these have been examined manually. All the original records are in British units and, while tabulated values have been converted to metric units, this is not possible for frequency histograms.

Preliminary comparison of more or less simultaneous readings in Lochs Morar and Shiel showed that a straight line would not describe the relationship over the entire range. Fluctuations in Loch Morar are condensed compared to those in Loch Shiel and the peaks and troughs do not always occur on the same day.

In order to generate a synthetic record for Loch Shiel, the twelve monthly means for 1978 were compared (Fig. 2.2) as were the relationships for peaks and troughs (Fig. 2.3).

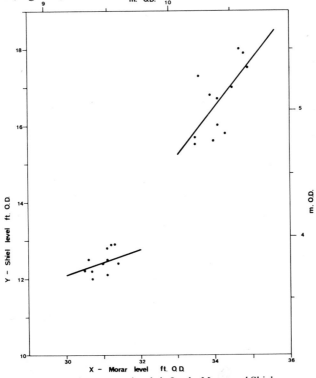

Fig. 2.3 Relation between the extreme levels in Lochs Morar and Shiel.
Peaks Y = 1.253X − 26.157 (feet units) Troughs Y = 0.331X + 2.163 (feet units)

Table 2.5 Seasonal water level data for the five lochs in (metres above Ordnance Survey Datum)

Site	Jan.	Feb.	March	April	May	June	July	Aug.	Sep.	Oct.	Nov.	Dec.
Loch Lomond at Balloch 1955–1964[1]												
Mean monthly level	8.00	7.95	7.78	7.70	7.61	7.36	7.42	7.56	7.72	7.85	7.89	8.12
Maximum level in month	8.74	9.17	8.44	8.43	8.27	7.76	7.70	8.29	8.50	8.73	8.66	8.75
Minimum level in month	7.34	7.11	7.09	7.16	7.15	7.07	7.08	6.93	6.96	7.16	7.48	7.51
Loch Lomond at Balmaha 1973–1977[2]												
Mean monthly level	8.41	8.30	7.95	7.82	7.75	7.64	7.51	7.57	7.69	7.87	8.03	8.04
Maximum level in month	9.41	9.27	8.34	8.40	8.23	8.16	7.87	7.96	8.37	8.68	9.10	9.03
Minimum level in month	7.61	7.79	7.72	7.41	7.34	7.37	7.27	7.32	7.14	7.31	7.27	7.59
Loch Awe 1964–1977												
Mean monthly level	36.43	36.32	36.23	36.08	36.08	36.03	36.08	36.01	36.08	36.30	36.43	36.47
Maximum level in month	37.73	37.49	37.52	37.12	37.25	37.25	37.12	36.82	37.09	37.80	37.46	37.67
Minimum level in month	35.72	35.81	35.66	35.30	35.30	35.30	35.45	35.36	35.08	35.08	35.70	35.70
Loch Ness at Fort Augustus 1948–1978												
Mean monthly level	15.60	15.85	15.84	15.80	15.75	15.66	15.67	15.71	15.75	15.83	15.87	15.93
Maximum level in month	16.72	17.40	16.85	16.26	16.11	15.96	16.29	16.29	16.57	16.59	16.57	17.48
Minimum level in month	15.32	15.42	15.40	15.29	15.29	15.35	15.27	15.25	15.25	15.42	15.50	15.50
Loch Morar 1950–1977												
Mean monthly level	10.22	10.16	10.11	10.10	10.05	10.00	10.09	10.07	10.10	10.20	10.22	10.23
Maximum level in month	10.79	10.91	10.91	10.55	10.52	10.64	10.73	10.62	10.82	10.90	10.79	10.96
Minimum level in month	9.66	9.66	9.51	9.48	9.33	9.39	9.27	9.33	9.36	9.48	9.51	9.48
Loch Shiel												
Mean monthly level	4.64	4.57	4.51	4.50	4.44	4.40	4.49	4.47	4.51	4.62	4.63	4.64
Maximum level in month	5.55	5.70	5.70	5.24	5.20	5.35	5.47	5.34	5.58	5.68	5.55	5.76
Minimum level in month	3.86	3.86	3.81	3.80	3.75	3.71	3.73	3.75	3.76	3.80	3.81	3.80

[1] I.e. prior to the construction of the River Leven Barrage.
[2] I.e. after the construction of the River Leven Barrage.

Results. Seasonal water level data for the five lochs are listed in Table 2.5. The highest and lowest monthly means and the greatest extremes being in italics.

The estimated data for Loch Shiel were converted directly from the long-term values for Loch Morar. None of the records are completely comparable since they do not refer to identical time periods and extreme values tend to increase with the length of record.

The highest monthly mean values and the highest recorded levels occur in December except for the short Loch Lomond record relating to the period after the construction of the River Leven Barrage. The differences between the two sets of Lomond records cannot be attributed totally to the construction of the barrage since higher than average levels in the period 1973–1977 were recorded on other lochs. The lowest monthly mean values are generally in June but the pattern is not so consistent and the lowest recorded levels are often later in the year, i.e. after a prolonged summer drought.

In comparing the magnitudes of the level changes, it is not possible to attach much importance to the synthetic results for Loch Shiel since they directly reflect the patterns at Loch Morar. The mean monthly variability, i.e. the highest monthly mean less the lowest monthly mean, and the extreme range are listed in Table 2.6. Even after the construction of the barrage, the mean seasonal variability at Loch Lomond is much greater than elsewhere but this is not true of the extreme ranges.

In Lochs Awe, Ness and Morar particularly, the differences between the

Table 2.6 Water level variability for the five lochs

Loch	Monthly variability		Extreme range	
	(a)	(b)	(a)	(b)
Lomond at Balloch	.76	7.5[1]/21.9[2]	2.24	22.2/64.5
Lomond at Balmaha	.89	8.8/25.6	2.27	22.5/65.3
Awe	.46	6.5	2.72	38.3
Ness	.27	1.0	2.23	8.3
Morar	.22	2.2	1.62	16.8
Shiel	.24	3.8	2.0	32.4

(a) is the vertical water level movement (m)

(b) is the horizontal distance (m) along the shore zone profile represented by these vertical movements. These are based on the mean shore zone slopes listed in table 2.2.

[1]) North basin.
[2]) South basin.

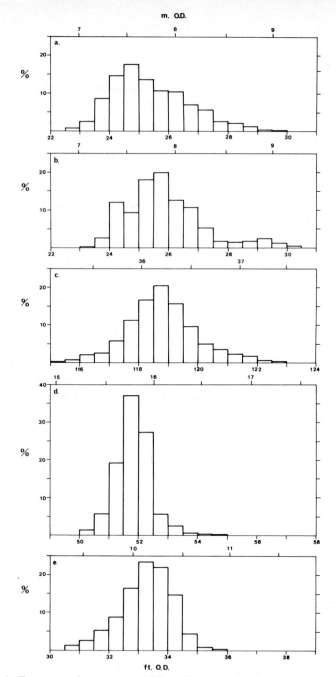

Fig. 2.4 Frequency of occurrence of different loch water levels.
a. Loch Lomond at Balloch (1955–1964), i.e. before the construction of the River Leven barrage.
b. Loch Lomond at Balmaha (1973–1977), i.e. after the construction of the River Leven barrage.
c. Loch Awe (1964–1977)   d. Loch Ness (1948–1978)   e. Loch Morar (1950–1977)

monthly and extreme patterns illustrate the inherent contradictions in any loch level control scheme. The desire to keep the loch level as steady as possible is reflected in fairly constant long-term monthly mean values. There may also be restrictions on the possible outflow rates, the actual outflow being less than might have occurred naturally at high water levels the converse occurring during droughts. These effects, which tend to increase the water level range, may be due to the restricted capacity of the outflow works themselves or to obligations to maintain flows below the lochs within certain ranges.

The frequency of occurrence of different water levels for the four lochs, i.e. excluding Loch Shiel, are illustrated in Fig. 2.4. The greater variability in Loch Lomond is again demonstrated.

Water levels covering the period when sampling at the five lochs was done, viz. November 1977 to October 1978, are illustrated in Fig. 2.5. The general pattern of the rise and fall of water level is somewhat similar for all five lochs although there are various differences which are consistent with the conclusions reached from the long-term records. Too much importance, however, should not be attached to conclusions drawn from these short records since there is no guarantee that the rainfall and runoff effects which generated the lochs' response are similar. The general impression is that the rate of rise in Loch Shiel particularly, but also in Loch Awe, is more rapid than in the others.

The features of the level fluctuations are not directly related to the loch surface area – Loch Lomond has both the largest area and the greatest level fluctuation – and the differences between the lochs are dependent on the form of the level – outflow relationship and the hydrological characteristics of the catchments.

Although there are differences in the behaviour of the five lochs, they are relatively small and it is unlikely that water level fluctuations, by themselves, are major causes of ecological differentiation. When, however, the fluctuations are linked with the mean shore zone slopes, as in Table 2.2, they are likely to be significant, but this reflects the properties of the shorelines not the water levels.

Radiation and temperature

Incoming radiation

The variation in latitute of the five lochs is small so that the incoming radiation at the top of the atmosphere is similar for all five. The actual energy supply at

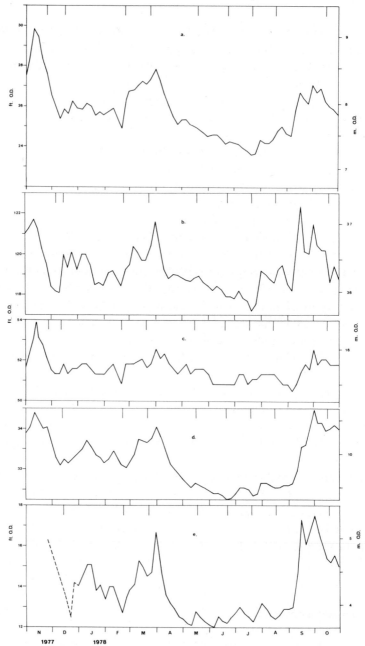

Fig. 2.5 Water levels for the five lochs during the period November 1977 to October 1978. Values plotted at five day intervals. The vertical lines are the dates on which shore zone samples were taken.

(a) Loch Lomond, (b) Loch Awe, (c) Loch Ness, (d) Loch Morar and (e) Loch Shiel.

the loch surface, i.e. the sum of the direct and diffuse radiation, is altered by the topography of the surrounding land and by the climatic conditions, primarily cloud cover. the question examined is whether the incoming radiation is sufficiently different for any of the lochs to have productivity differences due to variations in energy supply alone. The complex details of the energy balance are not considered.

Topographical analysis. For the standard sampling points on the five lochs, the angle of elevation to the skyline was measured on 1 : 50 000 maps at 108 horizontal intervals from 40° east of north to 320° and, compared with the solar altitude and azimuth. This was repeated for a point in the centre of the south basin of Loch Lomond.

Using a method described by the Building Research Station (Petherbridge 1966), the direct solar radiation on a clear day, i.e. when there is less than two eights cloud cover, can be calculated. The variation of direct radiation over the course of the day at various times of year are displayed in Fig. 2.6. The times when the sun is below the skyline are indicated.

The influence of topography is quite small although the influence of the orientation of the loch axis can be seen, particularly the very low loss in Loch Morar. It must be borne in mind that these comments refer to the standard sampling points, i.e. in the middle of the loch. The direct radiation cut off from certain parts of the shore may be considerable. The diffuse radiation, which is about one sixth of the direct at noon in midsummer and about two fifths of the direct at noon in midwinter, is even less affected by topography.

Effects of climate. The influence of cloud cover on incoming radiation is usually expressed as a comparison of the actual hours of bright sunshine with the theoretical. A number of empirical equations exist which can be used to convert records of bright sunshine, that quoted by Smith (1973) having been checked in Scottish conditions. Besides co-efficients related to astronomical conditions, this equation has the term $(0.3 + 0.7 \, n/N)$, n being the actual and N the theoretical hours of bright sunshine, which can be used to express relative radiation. Table 2.7 indicates the relative radiation for the limited number of weather stations in the area.

The variability is small and the high value at Tiree probably exaggerates the difference between Fort William and the seaward end of Lochs Morar and Shiel. Radiation differences, due to climate or topography, are not large enough to cause productivity differences due to energy supply alone.

Thermal characteristics. Temperature profiles were recorded using a

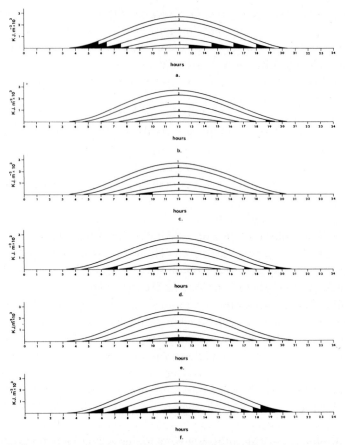

Fig. 2.6 Daily energy inputs from direct solar radiation for the five lochs.
Example dates are 1: 21st June; 2: 15th August; 3: 21st March/September; 4: 15th February; 5: 21st December.
The shaded areas indicate when there is no direct radiation reaching the standard sampling points because of shading by the surrounding hills.
(a) Loch Lomond (N. Basin), (b) Loch Lomond (S. Basin), (c) Loch Awe, (d) Loch Ness, (e) Loch Morar and (f) Loch Shiel.

Table 2.7   Variation in incoming radiation

Site	Actual average hours of sunshine per day	Relative radiation $(.3 + .7n/N)$	%
Fort William	2.75	.53	100
Inverness	3.37	.58	109
Tiree	3.97	.63	119
Helensburgh	3.29	.57	107

48

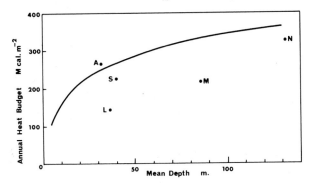

Fig. 2.7 Relation between mean depth and annual heat budget for the five lochs.
The curve is the regression of annual heat budget on mean depth given by Gorham (1964).

Weston and Stack temperature and oxygen probe with a 100 m of cable. There are occasional gaps in the record due to instrumental difficulties or severe weather conditions. The heat content of the five lochs was calculated by dividing the loch into a number of layers whose volume was estimated from volume-depth curves, multiplying these volumes by the mean temperature of the layer and summing the heat content of all the layers.

The annual heat budgets, i.e. the difference between the highest and lowest heat contents, are compared with mean depths on Fig. 2.7. Also shown is the regression of annual heat budget on mean depth for a large number of temperate lakes derived by Gorham (1964). Any conclusions based on a few measurements over a single year must be tentative but the values for Lochs Lomond and Morar do seem rather low. The reason is not altogether clear but is perhaps related to the morphometry of these lochs. The multiple basin character of Loch Lomond has already been referred to and the extreme maximum depth of Loch Morar suggests that its mean depth is somewhat greater than its modal or characteristic depth. There are, however, no major anomalies.

The thermal characteristics of the lochs are summarised in a series of charts (Fig. 2.8). The general form of the curves is the same for all so that, although detailed variations in, for example, the timing of the onset of stratification and overturn may be missed due to the infrequency of measurement, only Loch Ness can be seen to be noticeably different.

All are isothermal from late autumn to the spring with progressive development of stratified conditions from May to the following autumn. The very large volume of Loch Ness means that the onset of stratification is delayed as is the autumn overturn and the intensity of stratification, as measured by temperature gradient, is much reduced. All five, therefore, are warm, mono-

49

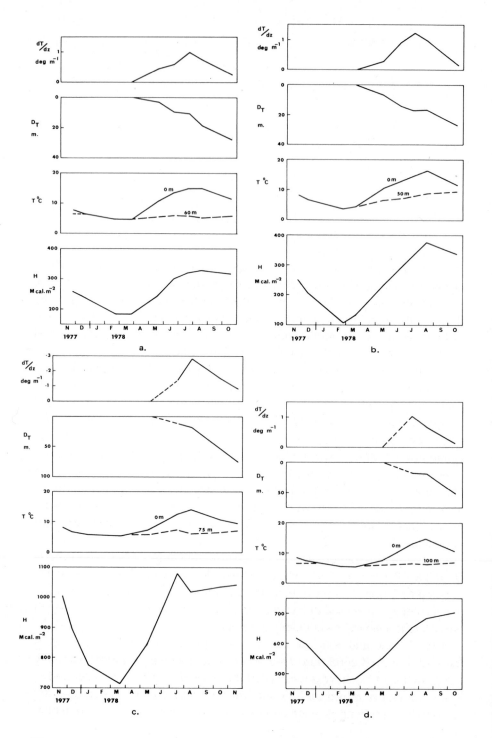

a.

b.

c.

d.

50

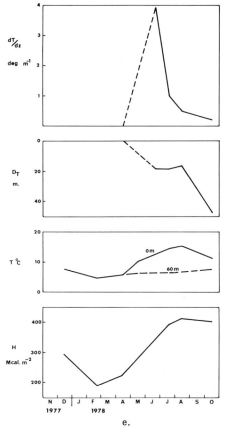

Fig. 2.8 Thermal characteristics of the five lochs.

H is the heat content, T is the water temperature, D_T is the depth of the maximum temperature gradient and dT/dz is the temperature gradient.

Sometimes only surface temperature was recorded in November 1977 and, thus, a vertical temperature difference may be unrecorded. In some cases, no significant temperature gradient could be detected when temperature differences between surface and depth were observed. These are shown as dashed lines in the upper graphs.

(a) Loch Lomond, (b) Loch Awe, (c) Loch Ness, (d) Loch Morar and (e) Loch Shiel

mictic lakes in the classification of Yoshimura (1936). The winter isothermal temperature of Loch Awe, however, does drop below 4 °C and no inverse stratification is observed.

The extent to which the observed temperature gradients are an effective barrier to mixing is not clear since this also depends on unmeasured velocity gradients. This is considered again later in relation to phytoplankton production where it is assumed that the existence of a temperature gradient of at least 0.5 deg m^{-1} implies the circulation of two separate layers. This is done simply to rank the lochs in terms of production efficiency and has no theoretical basis.

The external heat supply per unit area required to raise the temperature of a loch by one degree centigrade is simply the mean depth in units of Mcal m^{-2}, i.e. ranging from 32 Mcal m^{-2} for Loch Awe to 132 Mcal m^{-2} in the case of

51

Loch Ness. Referring to Fig. 2.6, the maximum, direct daily radiation on a clear day at mid-summer is less than 3×10^3 kJ m^{-2}, i.e. less than 12.5 Mcal m^{-2}, and about half this value at the equinoxes. Taking account of the fact that most of the incoming radiant energy is used up in evaporation or returned to the atmosphere through reflection, radiation and heat transfer plus the infrequency of cloud-free days in the area, it is obvious that the capacity to heat the lochs is limited.

The capacity to lose heat is similarly restricted. This, together with the windy conditions of a maritime climate, means that ice formation is limited to small areas of fringe ice and, occasionally, to extensive ice cover in the south basin of Loch Lomond. This also explains the absence of inverse stratification referred to above.

It is interesting to compare the rates of temperature change with the capacity of hydro-electric power stations located on the loch shores. In the case of Loch Awe, the Cruachan power station has a rated capacity of 400 MW. This implies that 204.5 h of continuous operation would be required to generate sufficient electrical energy to raise the temperature of Loch Awe by one degree. Given that normal operation is unlikely to be for more than 3 h/day, this is equivalent to at least two months of standard generating conditions.

Dissolved oxygen. At the same time as the temperature measurements, checks were made of the dissolved oxygen content of the water column. The results consistently showed all five lochs to be over 80% saturated (often approaching 100%) even during periods of stratification with no major irregularities around the discontinuity layer. In July, slight fluctuations around the thermocline were observed but, these were too small to be accurately determined by the dissolved oxygen probe being used.

Hydraulic conditions

Wind waves on the water surface

The application of modified oceanographic wave forecasting procedures to lakes has been considered by Smith & Sinclair (1972). The basis of the method is that, if the fetch and wind speed are known, then the wave characteristics in deep waters, i.e. where the lake bed has no influence, can be predicted with sufficient accuracy for ecological purposes. The resulting figures for wave height and length are the 'significant', i.e. they refer to the average of the highest third of all the waves and are a balance between giving too much emphasis to either small waves or the occasional large ones.

The fetch is not the greatest straight line distance over water but it can be calculated using an overlay on a large scale map so that the effects of lake width and irregular shorelines are taken into account. No detailed wind data for the five lochs are available. Birse & Robertson (1970) place the environs of all five in an exposure class corresponding to mean wind speeds ranging from 2.6–4.4 m s^{-1}. The actual wind speed over the water will be somewhat higher than this. The wave characteristics have been calculated using a wind speed of 10 m s^{-1}. Since Smith (1973) shows that a wind speed equal to twice the average value is exceeded between 15 and 20% of the time in Scotland, such a wind speed will generate appreciable yet reasonably frequent waves that characterise the wave climate of the lochs.

The results of applying the forecasting procedure at various points on the five lochs are displayed in a series of maps (Fig. 2.9). The obvious feature is that, because the lochs are narrow compared to their length (see Plate 2.2), the rate of wave growth downwind is small and that it is acceptable to specify a single typical wave for the whole loch. At this level of simplification, such a typical wave has been taken as having the average wave height and length for the particular loch.

In deep water, i.e. well away from the shore where the loch bed has no influence on the wave characteristics, the most important feature of water action is the thickness of the layer of water that is intensively mixed by wave action (see Plate 2.3). This can be defined by a depth equal to half the wave length. This results in mixed-layer depths ranging from just over 2.5 m in Loch Shiel to nearly 5 m in Loch Ness – in all cases small values compared to the mean depths of the lochs.

Wave action is ecologically much more important in the shore zone, i.e. where there is interaction between the wave characteristics and the loch bed. The limit of the shore zone, in this sense, can be defined as the point where the water depth is equal to half the wave-length. By combining this depth with the mean shore slopes as listed in Table 2.2 the characteristic shore zone length can be calculated. Within the shore zone, wave characteristics are transformed in a complex way but it is possible to estimate the approximate position of the breaker line since waves break, on average, when the water depth is four thirds of the wave-height (Hill 1962).

Between the shore zone limit and the breaker line, the hydraulic stress on the bed and thus the potential disturbance, increases. The occurrence of bed instability depends not only on the hydraulic stress but also on the resistance of the material forming the bed – normally defined by the particle size. In

53

general, the steeper the shore zone slope, the larger particles on the bed. On the land side of the breaker line, the bed is subject to the to and fro motion of the swash (see Plate 2.4). Plate 2.5 shows evidence of erosion on the shore of Loch Lomond due to the combined effects of wave action and recreational activity.

The features of the shore zone are illustrated in Fig. 2.10 and numerical values of the five lochs are noted in Table 2.8. It should be emphasised that the tabulated value of the shore zone length refers to a fixed water level. As the level changes, this shore zone length migrates up and down the shore. The absolute total length of bed potentially affected by wave action must take account of the horizontal distances represented by water level variability (Table 2.6). The extreme range of water level variability increases the affected length of shore zone by about 50% in the case of Lochs Ness and Morar and

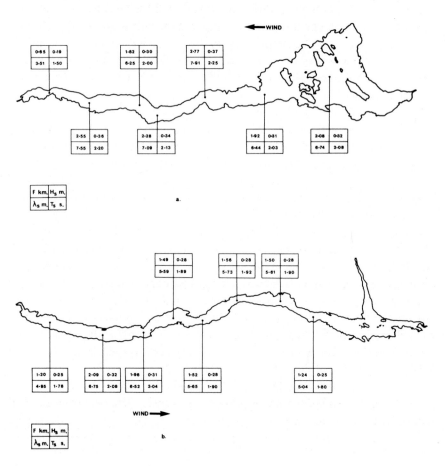

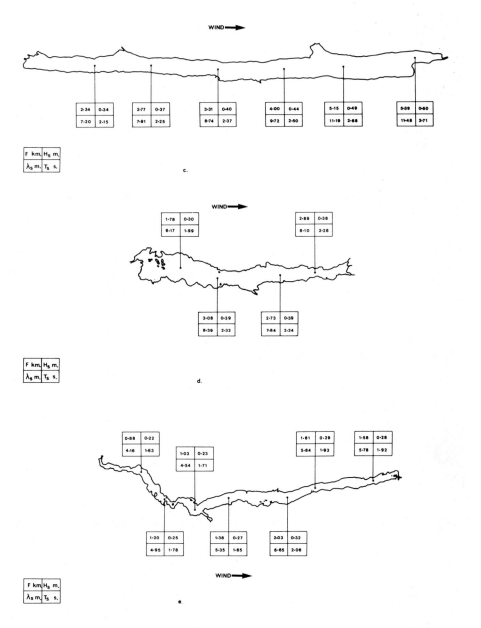

WIND ➡

| 2·34 | 0·34 | | 2·77 | 0·37 | | 3·31 | 0·40 | | 4·00 | 0·44 | | 5·15 | 0·49 | | 5·39 | 0·60 |
| 7·20 | 2·15 | | 7·91 | 2·25 | | 8·74 | 2·37 | | 9·72 | 2·50 | | 11·19 | 2·68 | | 11·48 | 2·71 |

| F km. | H$_s$ m. |
| λ$_s$ m. | T$_s$ s. |

c.

WIND ➡

| 1·78 | 0·30 | | 2·89 | 0·38 |
| 6·17 | 1·99 | | 8·10 | 2·28 |

| 3·08 | 0·39 | | 2·73 | 0·39 |
| 8·39 | 2·32 | | 7·84 | 2·24 |

| F km. | H$_s$ m. |
| λ$_s$ m. | T$_s$ s. |

d.

| 0·88 | 0·22 | | 1·61 | 0·29 | | 1·58 | 0·28 |
| 4·16 | 1·63 | | 5·84 | 1·93 | | 5·78 | 1·92 |

| 1·03 | 0·23 |
| 4·54 | 1·71 |

| 1·20 | 0·25 | | 1·38 | 0·27 | | 2·03 | 0·32 |
| 4·95 | 1·78 | | 5·35 | 1·85 | | 6·65 | 2·06 |

WIND ➡

| F km. | H$_s$ m. |
| λ$_s$ m. | T$_s$ s. |

e.

Fig. 2.9 Deep water wave characteristics for the five lochs.
The wind speed is 10 m s⁻¹ and the arrow indicates the wind direction. The upper left hand figure in the box is the effective fetch (km). The upper right figure is the significant wave height (m). The lower left hand figure is the wavelength (m). The lower right hand figure is the wave period (s).
(a) Loch Lomond, (b) Loch Awe, (c) Loch Ness, (d) Loch Morar and (e) Loch Shiel.

55

Plate 2.2 Oblique aerial view from the north east end of Loch Ness showing the longest freshwater fetch in Scotland. The storm beach on Plate 2.4 can be seen in the left of the picture. (Photo: Aerofilms Ltd.)

56

Plate 2.3 (above)   Storm conditions on Loch Ness. (Photo: A. A. Lyle.)
Plate 2.4 (below)   Waves breaking on the storm beach at the north east end of Loch Ness. The
location of the beach is shown on Plate 2.2. (Photo: A. A. Lyle.)

57

Plate 2.5 Evidence of shore erosion at Loch Lomond. (Photo: B. D. Smith.)

Table 2.8 Shore zone wave characteristics for the five lochs (see Fig. 2.10 for definitions)

Feature	Lomond		Awe	Ness	Morar	Shiel
	N. Basin	S. Basin				
×	9.91	28.78	14.09	3.71	9.92	15.97
A	3.25	3.25	2.87	4.69	3.82	2.66
B	32.21	93.44	40.44	17.40	37.89	42.48
C	0.41	0.41	0.37	0.56	0.51	0.34
D	4.06	11.80	5.21	2.08	5.06	5.43

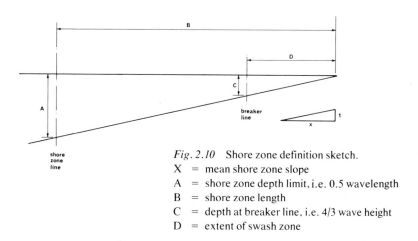

Fig. 2.10 Shore zone definition sketch.
X = mean shore zone slope
A = shore zone depth limit, i.e. 0.5 wavelength
B = shore zone length
C = depth at breaker line, i.e. 4/3 wave height
D = extent of swash zone

by around 80% for Lochs Shiel and Awe, Loch Lomond being intermediate. With the more common ranges of water level fluctuation, these percentages will be considerably reduced.

Although there are differences in the intensity of wave action in the five lochs, the ecological implications are, like the water level fluctuations, more directly related to the properties of the shore rather than to the wave characteristics themselves. Loch Ness, for example, is subject to the most intense wave action, yet, because of the steepness of its slopes, the length of shore zone affected is least.

Surface seiches

Analysis of water level oscillations was one of the major contributions in the studies on Scottish lochs earlier this century. See, for example, Chrystal (1904, 1905a & b). This early work was mainly concerned with the period of oscillation and with the breakdown of the primary or fundamental oscillation into a number of harmonics as with a musical instrument.

The period of a surface seiche is dependent solely on the geometry of the water body and, for the simple oscillation of a rectangular basin about the mid-point, can be easily calculated from Merian's formula (Hutchinson 1957). The formula is reasonably accurate when applied to basins of simple shape using the mean depth and effective length of the loch, the latter being estimated subjectively by examining plan and cross-sectional views of the loch. Calculations using Merian's formula (Table 2.9) are sufficient to indicate that the lochs, except for Loch Lomond, oscillate with fundamental periods of between 20 min and an hour.

Table 2.9   Seiche characteristics for the five lochs

Feature	Lomond	Awe	Ness	Morar	Shiel
Mean depth (m)	36.97	31.99	131.98	86.58	40.46
Effective length (km)	25.44	34.05	34.82	16.83	22.23
Period of primary surface seiche (min)	44.5	64.1	32.2	19.2	37.2
Set-up (wind of 10 m s^{-1}) (cm)	2.0	2.6	0.6	0.4	1.4

It is noticeable that, in the earlier work referred to above, no investigations seem to have been made of the oscillations of Loch Lomond. Loch Lomond, as already indicated, can be considered as a number of separate basins joined together and it is not uniform, its width varying irregularly, so that there are transverse flows as well as longitudinal oscillations. the resulting complexity is beyond the limit of analysis but the likelihood is that this will further increase the number of oscillations and reduce the period. The only experimental evidence, from a water level recorder in the upper end of the River Leven, maintained by the Clyde River Purification Board, certainly indicated a number of oscillations with much shorter periods than that calculated by Merian's formula (Table 2.9).

The greatest possible level difference over the length of the loch, i.e. the set up, can be calculated using the formula of Saville *et al.* (1962). The results, using the same wind speed as before, i.e. 10 m s^{-1}, are noted in Table 2.9. The main feature is that, since set up is inversely proportional to the depth, the level differences are small. Doubling the wind speed to 20 m s^{-1} (45 mph approx) results in a fourfold increase in the set-up, so that, even at high wind speeds, the maximum level difference in Loch Awe is little more than 10 cm. Seiche effects, therefore, are not a major influence on water levels in any of the five lochs.

Circulation and mixing

The main ecological interest in circulation and mixing concerns the homogeneity, or otherwise, of the lochs. There are various factors causing differentiation, such as variations in water depth and the chemical characteristics of inflow water, so that the use of single, biological sampling sites has to be justified. Experimental evidence relating to uniformity in some of the lochs is considered by George & Jones (1981), and the object here is to consider the existence of physical mixing mechanisms. Vertical differentiation is caused by temperature differences described earlier. There is no simple criterion based on temperature gradient which can be used to define when a loch will circulate in two separate layers since density effects must be considered in relation to the mechanical turbulence created by the wind-driven motion. It is likely, however, that for about a quarter of a year (July, August and September), the density gradients will be sufficient for the lochs to circulate vertically as two separate layers.

The idealised isothermal circulation in a long narrow loch of uniform depth assumes that approximately the upper third moves in the direction of the wind and that this transport of water is balanced by a slower moving current to the opposite direction over the lower two thirds of the depth (Smith 1979).

The actual circulation is complicated by the irregular geometry of the lochs and, at lower wind speeds, by deflections of the currents due to the effect of the earth's rotation. Such simplified models can be used, however, to calculate the order of magnitude of the total wind-driven current in the direction of the wind. Such a rate of water transport depends on the depth and is the difference between the drift current in the direction of wind and the gradient return current in the opposite direction. For the same wind conditions as used in the surface wave calculations, i.e. a wind speed of 10 m s^{-1}, the computed transports per unit width of loch have been calculated (Table 2.10).

The estimated total net forward transports, i.e. the transport per unit width multiplied by the mean breadth of the loch, are an order of magnitude greater than the inflow and outflow rates. Spatial differentiation, therefore, can, at these wind speeds, be rapidly reduced by wind-induced mixing. At lower winds and with the smaller effective depths when stratification occurs, the

Table 2.10   Total wind-driven transport in the five lochs for a wind speed of 10 m s^{-1}

Loch	Mean breadth	Transport per unit width	Total transport	Relative mixing time	Retention time/ relative mixing time
	km	m^2 s^{-1}	m^3 s^{-1}	days	
Lomond	1.95	0.548	1068.6	28.5	24.2
Awe	0.94	0.470	441.8	32.2	8.2
Ness	1.45	1.183	1715.3	50.3	20.4
Morar	1.42	1.003	1424.3	18.8	133.5
Shiel	0.70	0.597	417.9	21.9	22.8

transport rates are reduced so that spatial differences are observed at times. The occurrence of such spatial differences depends also on the general form of the loch and the proximity of major inflow streams to the outflow.

A measure of the scale of mixing is given by the relative mixing time in Table 2.10. This is the time for the given transport rate to move a volume of water equal to that of the loch itself. The range of times is relatively small, and, as with other features, the loch morphometry is probably more important than variations in the hydraulic conditions. it is noticeable, however, that Loch Morar, which has the shortest relative mixing time has the least detectable spatial variation (George & Jones 1981). The effect is increased when the relative mixing and retention times are compared.

Conclusions

This paper set out to consider how the interaction between the structure of the lochs and the external variables of rain, sun and wind influenced the physical environment within them. In broad terms the differences are not great. In particular, the external variables have a general similarity and what differences there are tend to reflect differences in morphometry. The features of shoreline profiles, for example, are more important than the effect of hydrological differences expressed as water level fluctuations or the influence of wind as it effects wave action while depth has more influence on the thermal structure than variations in incoming energy.

The broad generalisation, however, is not the whole story, particularly if the features of the catchment areas are taken into account. There is a tendency, admittedly, not always consistent, for those characteristics which favour biological productivity to be more pronounced in some lochs than others. Lochs Lomond and Awe, for example, have the lowest mean depths and the highest proportions of base rich geology and arable land within their catchments. The biology of the lochs may be more varied than an examination of the physical characteristics would suggest.

Table 2.11   Ranking of the physical characteristics of the five lochs

Feature	Lomond	Awe	Ness	Morar	Shiel
Surface area	1	3	2	4	5
Mean depth	4	5	1	2	3
Maximum depth	3	5	2	1	4
Volume	2	4	1	3	5
Shore zone slope	1	3	5	4	2
Inflow	3	2	1	5	4
Retention time	3	5	2	1	4
Mean seasonal range in water level	1	2	5	4	3
Absolute water level range	2	1	3	5	4
Surface temperature range	3	1	5	4	2
Thermocline depth	5	4	1	3	2
Degree of stratification	4	1	5	3	1
Wave height	3	4	1	2	5
Relative mixing time	3	2	1	5	4

Table 2.11 is a ranking of the main physical features of the five lochs, the highest numerical value of any attribute being ranked as one and the lowest as five. Loch Lomond, ranked first in terms of shore zone slope, has the flattest slopes. The mean seasonal water level range is the difference between the highest and lowest monthly mean water levels, the surface temperature range is the difference between the highest and lowest recorded temperatures, while the degree of stratification is an essentially subjective estimate based on the form of the temperature-depth profiles in July 1978, a high numerical value indicating a weakly defined thermocline.

Such rankings can be deceptive, particularly when the variations in the attributes are quantitative rather than qualitative differences in kind, but some features do emerge, the most obvious being the sheer size and slow response of Loch Ness. Lochs Lomond and Awe have a number of features in common, e.g. reduced mean and thermocline depths, greater water level and temperature variability and high inflows. Lochs Morar and Shiel are intermediate in mean depth and some other characteristics, Loch Shiel being perhaps more akin to Lochs Lomond and Awe while the great depth of Loch Morar makes it somewhat like Loch Ness.

63

Acknowledgements

We would like to thank a number of organisations for making available data on loch water levels – The Nature Conservancy Council and the Clyde River Purification Board (Loch Lomond), the British Waterways Board (Loch Ness) and the North of Scotland Hydro-Electric Board (Lochs Awe and Morar). Mr C. Macfarlane (Glenfinnan) read the staff gauge on Loch Shiel. We are particularly grateful to Mr R. M. Jarvis of the North of Scotland Hydro-Electric Board for the computer analysis of these levels and for other help and discussions.

We would also like to thank the Meteorological Office (Edinburgh) for assistance with climatic data and Mr T. Poodle of the Clyde River Purification Board for additional data on the River Leven (Loch Lomond).

References

Beran, M. A. & Gustard, A., 1977. A study into the low-flow characteristics of British rivers. J. Hydrol. 35: 147–157.

Birse, E. L. & Robertson, L., 1970. Assessment of climatic conditions in Scotland. Macaulay Institute: Soil Survey of Scotland.

Chrystal, G., 1904. Some results in the mathematical theory of seiches. Proc. R. Soc. Edinb. 25: 328–337.

Chrystal, G., 1905a. Some further results in the mathematical theory of seiches. Proc. R. Soc. Edinb. 25: 637–647.

Chrystal, G., 1905b. On the hydrodynamical theory of seiches. With a bibliographical sketch. Trans. R. Soc. Edinb. 41: 599–649.

George, D. G. & Jones, D. H., 1981. Spatial studies of the zooplankton of Scotland's largest lochs: Lomond, Awe, Ness, Morar and Shiel. In preparation.

Gorham, E., 1964. Morphometric control of annual heat budgets in temperate lakes. Limnol. Oceanogr. 9(4): 525–529.

Hill, M. N., 1962. The Sea. New York: Interscience Publishers.

Hutchinson, G. E., 1957. A treatise on limnology, New York: Wiley.

Maitland, P. S., 1981. The ecology of Scotland's largest lochs: Lomond, Awe, Ness, Morar and Shiel. Ed. P. S. Maitland. Introduction and catchment analysis. Chap. 1. This volume.

Meteorological Office, 1971. Climatological Memorandum No. 70. Frequencies of snow depths and days with snow lying at stations in Scotland for periods ending winter 1970/71. London: HMSO.

Meteorological Office, 1976a. Averages of temperature for the United Kingdom. 1941–70. London: HMSO.

Meteorological Office, 1976b. Averages of bright sunshine for the United Kingdom. 1941–70. London: HMSO.

Murray, J. & Pullar, L., 1910. Bathymetrical survey of the freshwater lochs of Scotland. Edinburgh: Challenger. Vols 1-6.

Petherbridge, P., 1962. Sunpath diagrams and overlays for solar heat gain calculations. Building Research Station, Current Papers. Series No. 39. Ministry of Technology.

Saville, T., McClendon, E. & Cochrane, A. L., 1962. Freeboard allowances for waves in inland reservoirs. Proc. Am. Soc. civ. Engrs. 88 WW2: 93–124.

Shellard, H. C., 1968. Tables of wind speed and direction over the United Kingdom. London: HMSO.

Smith, I. R., 1973. An assessment of winds at Loch Leven, Kinross. Weather Lond. 28(5): 202–209.

Smith, I. R., 1974. The structure and physical environment of Loch Leven, Scotland. Proc. R. Soc. Edinb. B 74: 81–100.

Smith, I. R. 1979. Hydraulic conditions in isothermal lakes. Freshwat. Biol. 9: 119–145.

Smith, I. R. & Sinclair, I. J., 1972. Deep water waves in lakes. Freshwat. Biol. 2: 387–399.

Yoshimura, S., 1936. A contribution to the knowledge of deep water temperatures of Japanese lakes. Jap. J. Astr. Geophys. 13: 61–120.

3.   Chemical characterisation – A one-year comparative study

A. E. Bailey-Watts & P. Duncan

Abstract

The chemical character of Lochs Lomond, Awe, Ness, Morar and Shiel, is described from analyses of the top 10 m of the water column, although data from depth samples and the main outflows are included. The study, based on nine samplings carried out between November 1977 and October 1978, shows the lochs to be very dilute (conductivities range from 29-41 μS cm^{-1} at 20 °C) and oligotrophic. Nevertheless, there are distinct chemical differences between these lochs; the comparatively alkaline Lochs Awe and Lomond (0.18 and 0.13 mEq alkalinity as $CaCO_3$ l^{-1} respectively) with Na:Ca weight ratios of 1.1, contrast with Lochs Morar and Shiel (0.06 and 0.04 alkalinity respectively) in each of which the Na:Ca ratio is 4.3. An attempt is made to relate these differences to features of the loch catchments. Relationships between nutrient levels and the phytoplankton populations of these waters, the results of a companion study, are discussed.

Introduction

This paper aims at a preliminary chemical characterisation of five large freshwater lochs lying between latitudes 56°05′ and 57°16′ N in Scotland: Lochs Lomond (north basin only), Awe, Ness, Morar and Shiel. The work described is based on nine sampling tours made between November 1977 and October 1978. Together with measurements of physical variables (see Smith *et al.* 1981b who include comments on dissolved oxygen in their section on vertical stratification) and analyses of the catchments of each loch (Maitland 1981), it comprises background environmental information to concurrent biological studies (see below). The limnological programme also assesses the possible

Plate 3.1a-b   Steep-sided basins of (a) Loch Lomond looking north, and (b) Loch Ness looking north-east.

68

Plate 3.1c-d Steep-sided basins of (c) Loch Morar looking south from Bracora, and (d) Loch Shiel looking south-east. (Photos: B. D. Smith, P. S. Maitland & A. E. Bailey-Watts.)

effects of existing pumped-storage hydro-electric schemes on Awe at Cruachan and Ness at Foyers (Maitland *et al.* 1981b).

Whilst a considerable amount of limnological work has been done on Loch Lomond by e.g. Slack (1957), Maitland (1966) and the Clyde River Purification Board (unpublished data), and a little on Loch Awe by Jenkin (1930) and Tippett (1978) the other waters have remained largely unstudied since the early bathymetrical and associated investigations of Murray & Pullar (1910). The classic studies by Wedderburn & Watson (1909) and Mortimer (1955) on currents and seiches respectively in Loch Ness are exceptions. Moreover, the more recent work by Glasgow University on Lomond, concentrates mainly on its south basin although Chapman (1965) and Maulood & Boney (1980) include chemical information on the north basin.

The lochs are generally long and narrow, with deep and steep-sided basins (Plate 3.1): Loch Morar ranks seventeenth deepest in the world (maximum depth 310 m – Murray & Pullar 1910). The waters are very dilute (0.2–0.5 mEq total cations. l^{-1}) – a feature which has given rise to certain analytical problems (see below). The lochs thus compare with oligotrophic lakes in other parts of the world e.g. Africa (Talling & Talling 1965), the Canadian Shield (Armstrong & Schindler 1971), Tasmania, (Croome & Tyler 1972), and certain of the most dilute of other British lakes, e.g. in the English Lake District (Mackereth 1957), Cheshire Meres (Gorham 1957a), and Cairngorms, Scotland (Gorham 1957b).

The sites

Fig. 3.1 shows the geographical location of the five lochs, and Table 3.1 physical features which may relate to their chemical character. Of all the freshwater bodies of the British Isles, Loch Morar has the greatest maximum depth, Loch Ness the greatest mean depth and volume, Loch Awe the greatest length, and Loch Lomond the greatest surface area (except for a few lochs in Ireland). All the lochs originated in ice-scoured trenches and are fjord-like in depth, width, length and proximity to the sea (Plate 3.2). But for narrow thresholds of rock and glacial debris, Lochs Shiel, Awe, Lomond and Morar would be sea lochs like many of their near neighbours on the Scottish west coast (Whittow 1977). The formation of Loch Ness is determined mainly by the Great Glen Fault, although scouring has also been important. In contrast, the formation of Loch Morar was mainly ice-mediated, and although perhaps facilitated by a submerged fault, it is not tectonic. Of Loch Lomond, the northern linear stretch (78 m mean depth) is the main concern of this

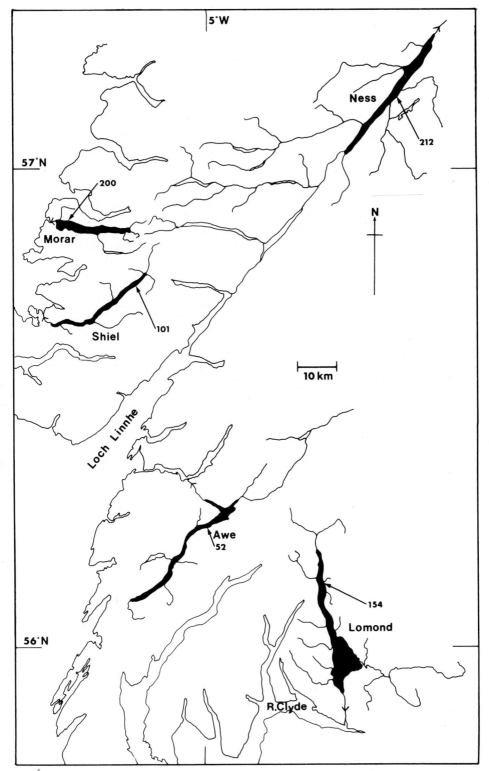

Fig. 3.1 Outline map of the five lochs studied, showing for each loch, the position of the open water sampling site (arrowed), the depth of water (in metres) at them, and the position of the main inflows and the outflow; (the direction of flow is indicated for the outflows only).

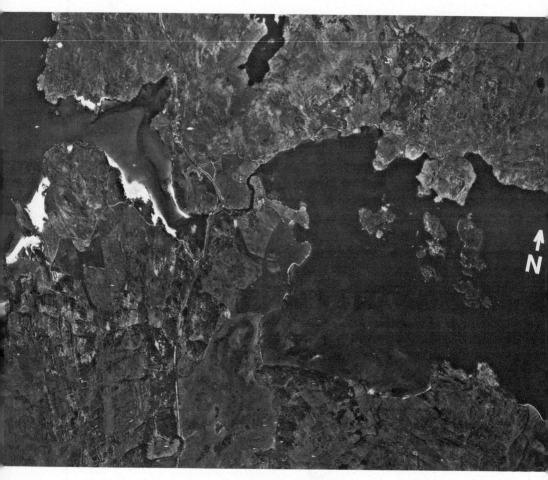

Plate 3.2 Vertical aerial view of the west end of Loch Morar showing its close proximity to the sea. Scale: 1 : 38 000. (Photo: Scottish Development Department.)

study; it differs markedly from the broader, shallow (23 m) southern basin and is largely separated from its influence by islands and outcrops of Old Red Sandstone associated with the Highland Boundary Fault. Physical features and catchment characteristics summarised in Table 3.1 are discussed more fully by Smith *et al.* (1981b) and Maitland (1981) respectively.

Edaphic features and morphometry of these waters are reflected in low population densities of phytoplankton (Bailey-Watts & Duncan 1981a), zoo-plankton (Maitland *et al.* 1981a) and zoobenthos (Smith *et al.* 1981a), and rather poor extents of macrophyte growth (West 1910, Spence e.g. 1964, 1967, and see Bailey-Watts & Duncan 1981b).

Table 3.1 Physical features and catchment characteristics of relevance to the study of water chemistry in the five lochs. (Data from Smith *et al.* 1981b, and Maitland 1981 who describe how the values were derived)

Feature	Lomond*	Awe	Ness	Morar	Shiel
Length (km)	19 (36)	41	39	19	28
Mean breadth (km)	1.00 (1.95)	0.94	1.45	1.42	0.70
Maximum depth (m)	190 (190)	94	230	310	128
Mean depth (m)	78 (37)	32	132	87	41
Catchment					
area (km²)	271 (781)	840	1775	168	248
% base rich rock	3 (45)	41	13	0	0
% arable land use	0.85 (17.0)	1.6	1.9	0.7	1.4
total population	627 (12 218)	777	3128	162	354
no. of sewage works discharges	1 (7)	2	8	0	0

* First values for Lomond refer to the oligotrophic northern basin with which this study is primarily concerned; values in parentheses refer to the whole loch.

Methods

Sampling

The lochs were sampled for water chemistry during each of nine 900-km week-long trips between November 1977 and October 1978. Most of the analyses reported here refer to samples taken at a single open water site in each loch (Fig. 3.1 shows their locations, and their depths). The open water station on Loch Awe is that referred to by Tippett (1978) as 'No. 1' or 'PS1' (Portsonachan). Although some spatial differences in chemical variables occur over the length of these lochs (George & Jones, in preparation, and unpublished data of the Clyde River Purification Board for Loch Lomond, and Tippett (1978) for Loch Awe) our results from single stations indicate the basic chemical character of each loch. Seasonal trends of some parameters are revealed although some changes have undoubtedly been masked by the relative infrequency of sampling (see related comments on the phytoplankton – Bailey-Watts & Duncan 1981a).

Six integrated tube samples of the top 10 m of the water column, and Ruttner water bottle collections from up to three discrete depths below this, were taken for chemical analysis. Profiles of temperature and dissolved oxygen throughout the water column were also taken at this station (Smith *et al.* 1981b). Brief reference to this information and data on phytoplankton (Bailey-Watts & Duncan 1981a) from the same points, is made in relation to the results discussed below. Dip samples from the outflows were also taken.

The lochs show a considerable range of water clarity and colour. Since this reflects to a large extent the differing contents of dissolved organic matter, light penetration data are included in this paper. As a routine on each sampling occasion these were obtained by Secchi disc – in this case a 25 cm black and white quartered disc. In addition on one occasion (autumn 1980) the spectral quality of the light field in each loch was measured. A submersible photo-cell and duplicate surface cell with diffusing opal and interchangeable Schott glass colour filters were used as described by Bindloss (1976) but with the recording instrument recently developed by Benham & George (1981).

Treatment and analysis of samples

On the boat, samples were stored in 5-l polythene containers. Immediately on return to the shore a 0.5-l subsample was transferred from each container to a polythene bottle and stored without freezing but as near 0 °C as possible, in ice-cooled insulated containers. These subsamples were finally transferred to a refrigerated store at the Institute of Terrestrial Ecology research station at Merlewood (Cumbria) until analysed. Two 2-l mixed aliquots of water from the tube samples were kept in polythene bottles for phosphate analysis only, with 6 ml of 2% mercuric chloride solution added as preservative.

Conductivity, pH, sodium, potassium, calcium, magnesium, Hazen colour and total organic carbon were determined using the instruments listed in Table 3.2. Continuous flow automatic colorimetry, in conjunction with the

Table 3.2 Instruments used in the analysis of conductivity, pH, alkalinity, cations, Hazen colour and total organic carbon

Analysis	Instrument used
Conductivity	Lock conductivity Bridge Type BC1 for routine measurements 1977–78 and the portable system described by Benham & George (1981) for measurements made in autumn 1980
pH	EIL pH Meter 7030 and 38B
Na, K	Corning EEL Model 150 Clinical Flame Photometer
Ca, Mg	Pye Unicam SP 1900 Atomic Absorption Spectrophotometer
Hazen colour	Lovibond 1000 Comparator
Total organic carbon	Carlo Erba 400/P Total Organic Carbon Analyser

Table 3.3 Some features of the analytical methods used in this study for nitrogen, phosphorus, silicon and chloride

Analysis	Chemical method used in conjunction with continuous flow automatic colorimetry	Limit of detection
NH_4–N	Colorimetric Indophenol Blue	100 μg l^{-1}
NO_2–N + NO_3–N	Naphthylamine-sulphonic acid	20 μg l^{-1}
PO_4–P	Molybdenum Blue	10 μg l^{-1}
SiO_3–Si	Molybdenum Blue	100 μg l^{-1}
Cl	Mercuric thiocyanate/ferric alum	500 μg l^{-1}

chemical methods shown in Table 3.3 was used to determine concentrations of ammonium-nitrogen, nitrate- plus nitrite-nitrogen, phosphate-phosphorus, silicate-silicon, and chloride. Alkalinity was determined by the standard method of titration, using screened methyl orange indicator. Measurements of total organic matter, and suspended and dissolved solids were discontinued after the first two sampling tours, in view of the low levels indicated and the method inaccuracies associated with them.

Results

The first part of this section concerns temporal changes in the general ionic environment of the surface waters, and the main differences between the five lochs in the levels and relative proportions of the major ions. Secondly, data on nutrients – nitrogen, phosphorus and silica – are presented together with information on water colour and light penetration. Still confining the comments to surface water chemistry, the relationships between the open water sites and the outflows are explored in the third part of this section. Finally, observations on the chemistry of deep water samples are summarised. Throughout, attempts are made to compare the lochs, to highlight their obvious distinguishing features and to rank them according to various chemical characteristics.

A few dissolved oxygen measurements were made during the study (Smith *et al.* 1981b). Vertical profiles show that even in summer the dissolved oxygen content is around 90% saturation from the surface to at least 100 m depth in the deepest lochs. Maulood & Boney (1980) report similar values for Loch Lomond north basin. This aspect of water chemistry is not discussed further.

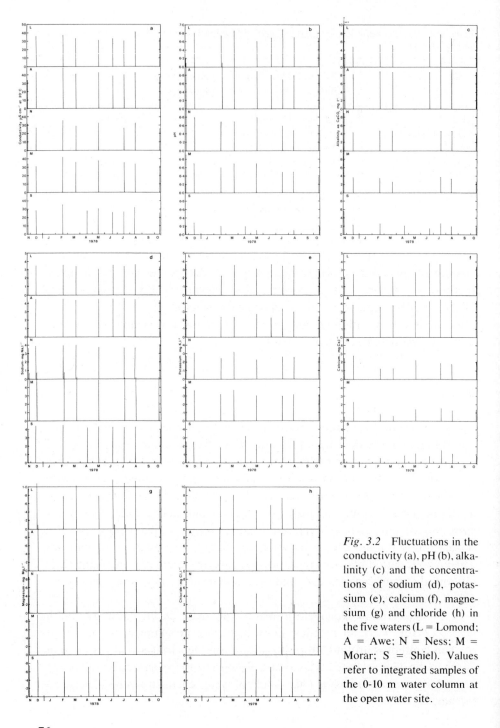

Fig. 3.2 Fluctuations in the conductivity (a), pH (b), alkalinity (c) and the concentrations of sodium (d), potassium (e), calcium (f), magnesium (g) and chloride (h) in the five waters (L = Lomond; A = Awe; N = Ness; M = Morar; S = Shiel). Values refer to integrated samples of the 0-10 m water column at the open water site.

76

General ionic environment of the surface waters

Temporal variations in the conductivity, pH and alkalinity, and levels of four major cations and chloride in the surface waters of the five lochs are presented in Fig. 3.2a-h. Average levels of all of these variables (Table 3.4) are lowest in Loch Shiel, except for sodium and chloride which are lowest in Loch Lomond. Highest average levels however are shared between Loch Morar (monovalent ions) Loch Awe (pH, alkalinity, conductivity and calcium) and Loch Lomond (magnesium). The overall range in conductivity (K_{20} of 29 to 41 μS cm^{-1}) indicates the extremely dilute nature of these waters; this is further confirmed (see below) by estimates of the total concentrations of ions.

Table 3.4 Lochs listed in sequences of mean annual levels (from lowest to highest) of various physico-chemical and chemical parameters

Conductivity μS cm^{-1} at 20 °C	K	Na	Ca	Mg	Alkalinity*	Cl	pH
←				mg l^{-1}			→
Shiel 29	Shiel 0.25	Lomond 3.45	Shiel 0.98	Shiel 0.75	Shiel 2.10	Lomond 6.52	Shiel 6.14
Ness 30	Ness & Awe 0.27	Ness 3.84	Morar 1.23	Ness 0.81	Morar 3.27	Shiel 7.36	Morar 6.63
Lomond 34		Shiel 4.24	Ness 1.99	Morar 0.88	Ness 4.48	Ness 7.53	Ness 6.70
Morar 35	Lomond 0.33	Awe 4.47	Lomond 3.00	Awe 0.99	Lomond 6.37	Awe 8.55	Lomond 6.78
Awe 41	Morar 0.34	Morar 5.34	Awe 4.01	Lomond 1.00	Awe 8.97	Morar 10.6	Awe 6.90

* As CaCO$_3$.

Temporal fluctuations observed in the concentrations of even 'conservative' elements such as potassium (Fig. 3.2e) could be due to a number of factors. These include variations due to the sampling and analytical techniques employed, (see Grimshaw in Allen *et al.* 1974) as well as real temporal variations. Errors associated with the analytical techniques are especially difficult to assess. However, results of many hundreds of analyses done during this study show that the combined variation attributable to the analytical techniques, and small-scale spatial differences around the boat during each collection of six tube (0–10 m) samples, is insignificant (authors' unpublished data). There are however errors of reproducibility in results from consecutive sampling tours due to analysis by different personnel (J. A. Parkinson & M. Grimshaw – personal communication).

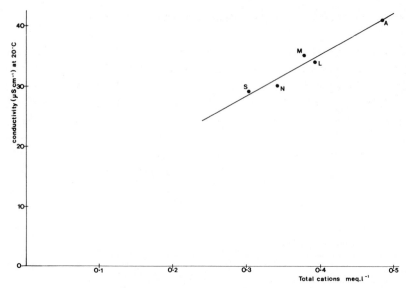

Fig. 3.3. The relationship between the mean values over the study period of conductivity and total cations in each loch (L = Lomond; A = Awe; N = Ness; M = Morar; S = Shiel). The line is fitted according to the regression equation y (conductivity at 20 °C in μS cm^{-1}) = 68.4 x (total cations in mEq l^{-1}) + 7.86 but has not been extended here to the intercept on the y axis. The correlation coefficient (r) for these points is 0.98.

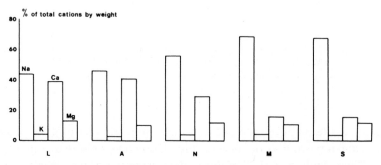

Fig. 3.4 Bar charts showing the percentage contributions of the four major cations to their combined total weight in Lomond (L), Awe (A), Ness (N), Morar (M) and Shiel (S) surface waters. Mean values for the study period have been used.

Despite these variations the relationship between mean values for the period of the study was extremely good; Fig. 3.3, relating conductivity to total cations, was plotted with deliberately expanded scales to accentuate differences between the lochs. The annual average equivalent conductivity (electrical conductivity related to total cations) of the five lochs is 90 μS cm^{-1} per mEq l^{-1}.

78

Conductivity values obtained with the field probe are all slightly higher than those shown in Fig. 3.3 which are the averages over the period of study of the values obtained with the laboratory meter i.e. Lomond 38, Awe 48, Ness 32, Morar 38 and Shiel 31 μS cm^{-1} at 20 °C. The position of Loch Morar in the sequence is determined largely by the high percentage contribution by sodium which is indicated in Fig. 3.4 by the higher mean annual sodium to calcium ratios in this loch and Loch Shiel.

A comparison of the ratios of monthly means of divalent: monovalent cations further highlights the individual character of each loch in spite of the

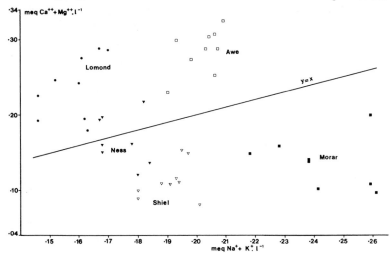

Fig. 3.5 The relationship between the combined concentrations (in mEq l^{-1}) of the divalent cations Ca and Mg, and those of the monovalent cations Na and K in the five lochs. Each point represents the mean value from six analyses (one for each of six 0-10 m tube samples) carried out on each of the 8 or 9 sampling occasions. Note the axes do not start at 0.

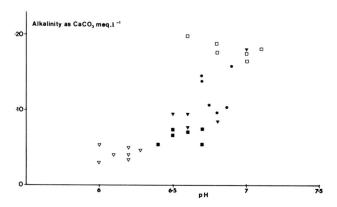

Fig. 3.6 As Fig. 3.5 for alkalinity (y axis) and pH (x axis, starting at 6 units) ● Lomond, □ Awe, ▼ Ness, ■ Morar, ▽ Shiel.

79

scatter of points due to temporal and other variability referred to above. Fig. 3.5 relates the sums of the monthly means, in equivalents per litre, of calcium and magnesium concentrations, to those of sodium and potassium; deliberately expanded axes have again been used. The loch sequence obtained from these ratios, runs from the 'low' pair Lochs Morar and Shiel (0.55 and 0.58 mean annual values respectively) to the 'high' pair Lochs Awe and Lomond (1.41 and 1.47 respectively) with Loch Ness in the middle (0.96). A similar sequence is exhibited by the data for both alkalinity and pH which are reasonably well related (Fig. 3.6). Levels of cations, chloride, pH and alkalinity similar to those given above have been obtained by the Freshwater Fisheries Laboratory of DAFS at Pitlochry (L. A. Caines – personal communication) from occasional analyses of these lochs in the past, and by Tippett (1978) during his 1975–76 study of Loch Awe.

Nutrients and water colour of the surface waters

The ammonium nitrogen concentrations exceeded the limit of detection (100 μg N l^{-1}) only once during the year's sampling: in October 1978 in Lochs Awe and Shiel. Values for the oxidised forms of inorganic nitrogen (nitrate plus nitrite – Fig. 3.7) remained lower than 100 ug N l^{-1} in Lochs Shiel and Morar

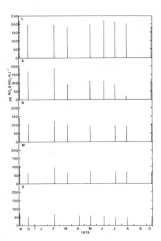
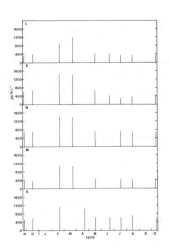
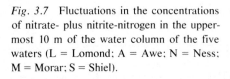

Fig. 3.7 Fluctuations in the concentrations of nitrate- plus nitrite-nitrogen in the uppermost 10 m of the water column of the five waters (L = Lomond; A = Awe; N = Ness; M = Morar; S = Shiel).

Fig. 3.8 As Fig. 3.7 for silicate-silicon concentrations.

throughout the period, but exceeded this value on 4 occasions in Loch Ness, 6 in Loch Awe and was consistently above it, sometimes exceeding 200 μg l⁻¹, in Loch Lomond. Dissolved orthophosphate concentrations in all lochs remained below the limit of detection (10 μg P l⁻¹) of the analytical method employed. The paucity of nutrients in Loch Awe and Loch Lomond has been reported by Tippett (1978) and Maulood & Boney (1980) respectively; these authors also show that levels of orthophosphate rarely exceed 1 ug P l⁻¹ in these two lochs which in the present study may be considered the richest of the five waters.Dissolved silicon exhibited marked seasonal fluctuations (Fig. 3.8), with spring levels exceeding 1 mg Si l⁻¹ in all the lochs. Decreases from these maxima to lower levels in summer coincided only broadly with the periods of diatom growth in the lochs (Bailey-Watts & Duncan 1981a). Comparable silica values were obtained by Tippett (1978) for Loch Awe and Maulood & Boney (1980) for Loch Lomond.

Water colour varied between 5 and 10 Hazen units in the clear Lochs Shiel and Morar. Similar values were recorded for Loch Lomond, although a few samples were recorded as 15. Contrastingly the brown waters of Lochs Awe and Ness rarely yielded values of less than 15. Total organic carbon concentrations in the latter pair of lochs were correspondingly higher than in the others – i.e. around 3 mg C l⁻¹ compared with 2.

It is to be expected that the clear and stained waters can also be distinguished by their Secchi disc transparencies, and light attenuation properties. Fig. 3.9 shows the contrasting ranges in Secchi readings taken at each loch and the differences in the rates of absorbance of light of various wavelengths. In all five lochs the light registered with the blue filter – Schott type BG12 with an optical mid-point of 460 nm – is that attenuated most rapidly. However the depth at which this light is reduced to 1% of the surface intensity is smallest in the brown waters of Lochs Awe and Ness – 1.2 and 1.3 m respectively. In Loch Morar the value was 3.4 m and in both Lochs Shiel and Lomond an intermediate value of 2.2 m was recorded. The waveband of light showing the minimum extinction however was not the same in each loch. The stained waters of both Lochs Awe and Ness were the most transmissive in light registered with the red filter (Schott type RG1 – optical mid-point 630 nm). This is the case in other brown-water lakes e.g. L. Pääjärvi in southern Finland (Jones & Ilmavirta 1978). In the two Scottish lochs the 1% depth for this light was similar – 4.3 m in Loch Awe and 4.7 m in Loch Ness. Contrastingly in Lochs Morar and Shiel orange light registered with the Schott type OG2 + BG18 filter combination (optical mid point 590 nm) penetrated the most, being reduced to 1% of the surface intensity at 9.5 m in Loch Morar and 8.4 in Loch Shiel. In Loch Lomond green light (registered with the Schott

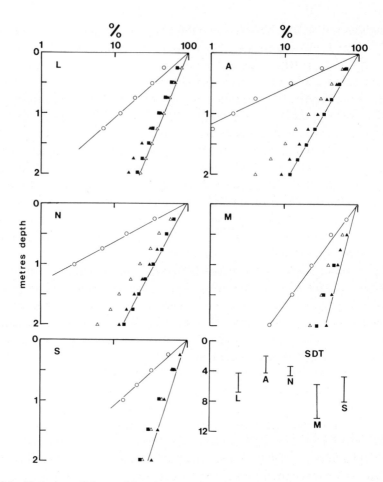

Fig. 3.9 Underwater light conditions in Lochs Lomond (L), Awe (A), Ness (N), Morar (M) and Shiel (S). Penetration of blue (O), red (■), orange (▲) and green (△) light expressed as changes in percentages (%) of surface values. Lines have been fitted (by eye) only to the points referring to the colours showing maximum and minimum attenuation. Bottom right-hand figure shows the ranges in Secchi disc transparency (SDT) recorded for each loch.

filter type VG9 with an optical mid point of 540 nm) was the least rapidly attenuated, its intensity being reduced to 1% of the surface value at 6.25 m. Of light in the green, blue, red and near ultra-violet bands, that in the green band was found by Talling (1971) to penetrate the furthest into a wide trophic range of English Lake District lakes. The similarity between the slopes of attenuation of green and red light in Loch Lomond (Fig. 3.9) however, resembles more the spectral properties of Talling's eutrophic lake type (e.g. Blelham Tarn,

Loweswater) than of either the oligo- or meso-trophic types in which red light is cut out by relatively much shallower depths than the green.

Chemistry of outflows and depth samples compared to surface waters

For most of the chemical parameters measured, there were no statistically significant differences in concentration between the surface water, and either the main outflow or the deeper water samples. Exceptions to this rule for each loch are as follows: In Loch Lomond most values recorded for the main outflow, which lies at the richer (mesotrophic – see Ratcliffe 1977) southern end of the loch, exceeded those of the northern basin, although dissolved silicon concentrations in the summer were somewhat lower in the outflow. Silicon, chloride and calcium showed slight increases with depth in the summer. In Awe one of the few notable exceptions to the general similarity in chemical content of surface, depth and outflow waters was the summer increase of silicon with depth which was also observed in Loch Ness and to a lesser extent in Loch Morar. Concentrations of silicon and nitrate-

Table 3.5 Dissolved silicon concentrations (mg Si l^{-1}) at the surface of the open water station (T, range of six 0–10 m tube samples) and the outflow (F, range of two dip samples) of each loch. Where only one value is entered there was no variation between samples

Sampling		Lomond	Awe	Ness	Morar	Shiel
Nov. 1977	T	0.35	0.58–0.60	0.69	0.37–0.43	0.54–0.57
	F	—	—	—	0.41	—
Dec.	T	0.35–0.40	0.60–0.70	0.65–0.75	0.40–0.44	0.56–0.60
	F	0.60–0.65	0.65–0.68	0.65	0.42–0.46	0.65–0.70
Feb. 1978	T	0.88–0.92	1.30–1.70	1.30–1.50	0.98–1.10	1.00–1.30
	F	1.70–1.80	1.30–1.80	1.30–1.40	1.00–1.10	1.20
March*	T	1.10–1.20	1.40–1.50	1.30–1.50	0.88–1.20	1.00–1.10
	F	1.40	1.40–1.60	1.40	0.90–1.00	1.10–1.20
May	T	0.42–0.43	0.64–0.65	0.71–0.72	0.44	0.60–0.61
	F	0.19–0.20	0.61–0.63	0.71	0.43	0.56
June	T	0.39–0.40	0.35–0.41	—	—	0.63–0.65
	F	0.48–0.49	0.64–0.70	—	—	0.64–0.65
July	T	0.33–0.37	0.24–0.26	0.71–0.74	0.41–0.42	0.63–0.64
	F	0.14	0.39–0.40	0.65	0.41–0.42	0.57
Aug.	T	0.33–0.34	0.33–0.34	0.65–0.68	0.37–0.39	0.70–0.72
	F	0.20	0.43–0.44	0.64–0.65	0.36–0.37	0.64
Oct.	T	0.34	0.48–0.50	0.67–0.70	0.40	0.58
	F	0.37–0.38	0.55–0.56	0.63–0.66	0.42–0.43	0.60–0.61

* April for Loch Shiel.

plus nitrite-nitrogen were also higher in the main outflow of Loch Awe in summer. In Loch Shiel silicon showed a tendency to increase slightly with depth, and rather lower concentrations were found in the outflow. In contrast to the other lochs except Loch Lomond, four cations and chloride as well as conductivity, were higher at the outflow than in the open water at our sampling point. The most marked variation was shown by dissolved silicon concentrations; Table 3.5 and Fig. 3.10 summarise the differences between surface water and outflows, and surface water and depths respectively.

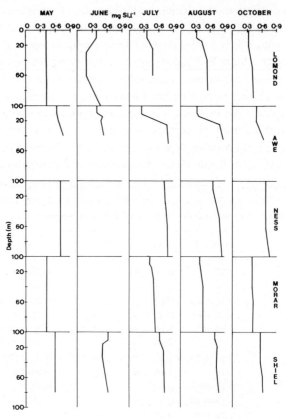

Fig. 3.10 Variation with depth in the concentration of dissolved silicon of the five lochs (L = Lomond; A = Awe; N = Ness; M = Morar; S = Shiel). Spring, summer and autumn periods are included.

Discussion

Firstly in this section differences in chemical composition of the lochs are considered. Secondly, in view of the fundamental importance of these data to

84

an understanding of the biology of the lochs, a few aspects of chemistry-phytoplankton relations are discussed. Despite similarities in morphometry of the lochs, an investigation of their chemistry reveals obvious differences. Mackereth (1957) and Gorham (1957a, b) demonstrated marked variation in the ionic composition of waters in geographical regions far smaller than the one encompassing these five large lochs, but their studies included a large number of small pools, tarns and lochans.

Measured conductivity values are in keeping with conductance estimates calculated by summing the contributions (equivalent conductances) of the major ions present (see Table 4.1 of Golterman 1969). The relationship between average annual values of conductivity and total cation content for the five lochs, 90 μS cm^{-1} per mEq l^{-1} (for 20 °C), is similar to Talling & Talling's (1965) calculated average equivalent conductivity, 85 (as μmho per mEq l^{-1}), for African waters. However, although a wide variety of lakes is included in their review, the principal ions are sodium with bicarbonate.

These five Scottish waters comprise at least two distinct chemical types. For present purposes the terms 'alkaline' and 'saline' are used to describe them. Since each type is associated with a different set of ions, different equivalent conductivities may have been expected (Hutchinson 1957, Mackereth et al. 1978). Lochs Awe and Lomond are of the alkaline type; they exhibit the highest alkalinity and pH values in the series, and the highest calcium (and magnesium) levels expressed both in absolute terms and as a percentage of total cationic weight. Whilst in one of these lochs – Loch Awe – the second highest level of sodium was found, the pair contrast with the saline waters, Lochs Morar and Shiel, where this element is the dominant cation. In Loch Morar the highest absolute values of sodium were recorded, and in both Lochs Morar and Shiel it is the major contributor to the total cationic weight.

The sodium to calcium ration (by weight) further illustates the difference between the Awe/Lomond and Morar/Shiel water types; in the former this ratio approximates 1.1, in the latter 4.3 with an intermediate value of 1.9 in Loch Ness. These differences may result from sodium inputs from marine influences on the one hand and the solution of bases from catchment bedrock on the other.

In their higher inorganic nitrogen levels (and phytoplankton densities – Bailey-Watts & Duncan 1981a) the 'rich' Lochs Lomond and Awe also contrast with the three other – ultra-oligotrophic – lochs. For Loch Awe indices of potential nutrient loading levels – e.g. area of catchment and the area of arable land and number of people in it related to loch volume also are high. Similar indices calculated for Loch Lomond north basin however are very low, being more comparable with the values obtained for the poorer lochs; for this reason, the possible influence of water from the much richer (even cf. Loch Awe) south basin of Loch Lomond cannot be ignored (Maitland *et al*. 1981c).

Bailey-Watts & Duncan (1981a) describe the different densities of phytoplankton populations occurring in these lochs. Algal abundance increased broadly with productive potential of the waters, as indicated by alkalinity and nutrient levels (see also Lund 1957). However the relationship between algal quality and/or abundance, and chemical concentrations varied considerably with the parameters used. Whilst this situation results in part from our incomplete understanding of the chemical-biological interactions involved, the relative infrequency of sampling of this study is undoubtedly another important contributory factor. This has, above all, prevented an analysis of short-term fluctuations in chemical and biological parameters of the lochs.

Our limited data on N and P suggest that these elements vary little with depth even during the summer period of thermal stratification. Similar findings by Jenkin (1930) and Tippett (1978) on Loch Awe led them to suggest that little release of nutrients from the sediments takes place. This is largely due to the lack of deoxygenation in the hypolimnetic waters. In the suspected absence of a large release of silicon from the sediments and although its biological uptake is limited in the depths, this element may be renewed at least partially by decomposition of diatom frustules during their gradual descent through the water column. This could explain the increases in silicon concentration with depth (Fig. 3.10). This is confirmed by microscopic evidence of extremely 'thin' diatom frustules – especially of *Asterionella* – which have been observed. No such partly dissolved fragments have been noticed in other studies of the water column of the shallow Loch Leven (Bailey-Watts 1978), although evidence of diatom dissolution in its sediments was obtained from field and laboratory studies (Bailey-Watts 1976). Jenkin (1930) postulated the same mechanism for Awe. More recent findings of marine studies suggest that the faster dissolution of delicate forms can result in a sedimentary cell assemblage that is quite different from the living plankton at the surface (Calvert 1966, Round 1967).

In conclusion, the colour of these waters deserves brief comment. It is debatable whether the term 'dystrophic' should be used to describe the more humic-stained Lochs Awe and Ness (see e.g. Hutchinson 1957). However Järnefelt (1958), in discussing a wide range of humus-coloured water bodies in Fenno-Scandinavia, considers the dystrophic type supplementary to, rather than comparable with, the oligotrophic and eutrophic series (Naumann 1919). Furthermore Jenkin (1930) concluded that Loch Awe was oligotrophic, though differing from other 'Alpine lakes' in its colour and transparency. She

86

also drew a distinction between this loch and the true dystrophic types in that the allochthonous matter content of the former (cf. also Loch Ness) is not high enough to reduce the oxygen content to the considerable extent characteristic of the latter.

Acknowledgements

This research was done under contract to the North of Scotland Hydro-Electric Board and the Nature Conservancy Council (I.T.E. Projects 546 and 545). We thank Mr A. J. Rosie for help in planning the sampling programme and in sharing with one of us (P.D.) all aspects of the field work. The I.T.E. Chemical Section at Merlewood, in particular Mr S. E. Allen, Mr J. A. Parkinson and Mr M. Grimshaw, are thanked for the considerable amount of analytical work done, and their constructive interest during all stages of the study and our preparation of this paper. The Director and staff of the Clyde River Purification Board are thanked for allowing us to consult their routine analytical results for Loch Lomond. Mr A. Kirika is gratefully acknowledged for help with data analysis, Ms G. M. Dennis for preparing the final versions of the figures, and Mrs M. S. Wilson for producing the typescript. We thank Dr P. S. Maitland, Mr I. R. Smith, Ms B. D. Smith, Mr A. A. Lyle and Mr A. J. Rosie, for many fruitful discussions throughout the study. Finally our gratitude to Mr L. A. Caines is expressed for his helpful criticism of this paper at an earlier stage.

References

Allen, S. E., Grimshaw, H. M., Parkinson, J. A. & Quarmby, C., 1974. Chemical analysis of ecological materials. Oxford: Blackwell.

Armstrong, F. A. J. & Schindler, D. W., 1971. Preliminary chemical characterisation of waters in the Experimental Lakes Area, north western Ontario. J. Fish. Res. Bd. Can. 28: 171–187.

Bailey-Watts, A. E., 1976. Planktonic diatoms and silica in Loch Leven, Kinross, Scotland: a one-month silica budget. Freshwat. Biol. 6: 203–213.

Bailey-Watts, A. E., 1978. A nine-year study of the phytoplankton of the eutrophic and non-stratifying Loch Leven, Kinross, Scotland. J. Ecol. 66: 741–771.

Bailey-Watts, A. E. & Duncan, P., 1981a. The ecology of Scotland's largest lochs: Lomond, Awe, Ness, Morar and Shiel. Ed. P. S. Maitland. The phytoplankton. Chap. 4. This volume.

Bailey-Watts, A. E. & Duncan, P., 1981b. The ecology of Scotland's largest lochs: Lomond, Awe, Ness, Morar and Shiel. Ed. P. S. Maitland. A review of macrophyte studies. Chap. 5. This volume.

Benham, D. G. & George, D. G., 1981. A portable system for measuring water temperature, conductivity, dissolved oxygen, light attenuation and depth of sampling. Freshwat. Biol. In press.

Bindloss, M. E., 1976. The light-climate of Loch Leven, a shallow Scottish lake, in relation to primary production by phytoplankton. Freshwat. Biol. 6: 501–518.

Calvert, S. E., 1966. The accumulation of diatomaceous silica in the sediments of the Gulf of California. Bull. geol. Soc. Am. 77: 569–696.

Chapman, M. A., 1965. Ecological studies on the zooplankton of Loch Lomond. Ph. D. Thesis. University of Glasgow.

Croome, R. L. & Tyler, P. A., 1972. Physical and chemical limnology of Lake Leake and Tooms Lake, Tasmania. Arch. Hydrobiol. 70: 341–354.

George, D. G. & Jones, D. H., 1981. Spatial studies of the zooplankton of Scotland's largest lochs: Lomond, Awe, Ness, Morar and Shiel. In preparation.

Golterman, H. L., 1969. Methods for chemical analysis of fresh waters (I.B.P. Handbook No. 8). Oxford: Blackwell.

Gorham, E., 1957a. The ionic composition of some lowland lake waters from Cheshire, England. Limnol. Oceanogr. 2: 22–27.

Gorham, E., 1957b. The chemical composition of some natural waters in the Cairngorm-Strathspey district of Scotland. Limnol. Oceanogr. 2: 143–154.

Hutchinson, G. E., 1957. A treatise on limnology. 1. New York: Wiley.

Järnefelt, H., 1958. On the typology of the northern lakes. Verh. int. Verein. theor. angew. Limnol. 13: 228–235.

Jenkin, P. M., 1930. A preliminary limnological survey of Loch Awe (Argyllshire) Part 1: An investigation of some physical and chemical conditions in the loch and experiments on photosynthesis at various depths. Int. Revue ges. Hydrobiol. Hydrogr. 24: 24–46.

Jones, R. I. & Ilmavirta, V., 1978. Vertical and seasonal variation of phytoplankton photosynthesis in a brown-water lake with winter ice cover. Freshwat. Biol. 8: 561–572.

Lund, J. W. G., 1957. Chemical analysis in ecology illustrated from Lake District tarns and lakes. 2. Algal differences. Proc. Linn. Soc. Lond. 167: 165–171.

Mackereth, F. J. H., 1957. Chemical analysis in ecology illustrated from Lake District tarns and lakes. 1. Chemical analysis. Proc. Linn. Soc. Lond. 167: 159–164.

Mackereth, F. J. H., Heron, J. & Talling, J. F., 1978. Water analysis: some revised methods for limnologists. Freshwater Biological Association Scientific Publications No. 36.

Maitland, P. S., 1966. Studies on Loch Lomond 2. Glasgow: Blackie.

Maitland, P. S., 1981. The ecology of Scotland's largest lochs: Lomond, Awe, Ness, Morar and Shiel. Ed. P. S. Maitland. Introduction and catchment analyses. Chap. 1. This volume.

Maitland, P. S., Smith, B. D. & Dennis, G. M., 1981a. The ecology of Scotland's largest lochs: Lomond, Awe, Ness, Morar and Shiel. Ed. P. S. Maitland. The crustacean zooplankton. Chap. 6. This volume.

Maitland, P. S., Smith, I. R., Bailey-Watts, A. E., George, D. G., Lyle, A. A., Smith, B. D., Duncan, P., Rosie, A. J., Dennis, G. M. & Carr, M. J., 1981b. Twenty-four hour studies of the effects of pumped storage power stations on water and plankton in Loch Awe (Cruachan) and Loch Ness (Foyers), Scotland. In preparation.

Maitland, P. S., Smith, I. R., Bailey-Watts, A. E., Smith, B. D. & Lyle, A. A., 1981c. The ecology of Scotland's largest lochs: Lomond, Awe, Ness, Morar and Shiel. Ed. P. S. Maitland. Comparisons and synthesis. Chap. 10. This volume.

Maulood, B. D. & Boney, A. D., 1980. A seasonal and ecological study of the phytoplankton of Loch Lomond. Hydrobiologica. 71: 239–259.

Mortimer, C. H., 1955. Some effects of the earth's rotation on water movement in stratified lakes. Verh. int. Verein. theor. angew. Limnol. 12: 66–77.

88

Murray, J. & Pullar, L., 1910. Bathymetrical survey of the freshwater lochs of Scotland. Edinburgh: Challenger. Vols. 1-6.

Naumann, E., 1919. Några synpunkter angående limnoplanktons okologi med sarskild hansyn till fytoplankton Sv. Bot. Tidskr. 13: 129–163.

Ratcliffe, D. A., 1977. A nature conservation review. 1. Cambridge University Press.

Round, F. E., 1967. The phytoplankton of the Gulf of California. Part. I. Its composition, distribution and contribution to the sediments. J. exp. mar. Biol. Ecol. 1: 76–97.

Slack, H. D., 1957. Studies on Loch Lomond. 1. Glasgow: Blackie.

Smith, B. D., Maitland, P. S., Young, M. R. & Carr, M. J., 1981a. The ecology of Scotland's largest lochs: Lomond, Awe, Ness, Morar and Shiel. Ed. P. S. Maitland. The littoral zoobenthos. Chap. 7. This volume.

Smith, I. R., Lyle, A. A. & Rosie, A. J., 1981b. The ecology of Scotland's largest lochs: Lomond, Awe, Ness, Morar and Shiel. Ed. P. S. Maitland. Comparative physical limnology. Chap. 2. This volume.

Spence, D. H. N., 1964. The macrophytic vegetation of lochs, swamps and associated fens. In: The vegetation of Scotland. Ed. J. H. Burnett. Edinburgh: Oliver & Boyd. 306–425.

Spence, D. H. N., 1967. Factors controlling the distribution of freshwater macrophytes with particular reference to the lochs of Scotland. J. Ecol. 55: 147–170.

Talling, J. F., 1971. The underwater light climate as a controlling factor in the production ecology of freshwater phytoplankton. Mitt. Internat. Verein. theor. angew. Limnol. 19: 214–243.

Talling, J. F. & Talling, I. B., 1965. The chemical composition of African lake waters. Int. Rev. ges. Hydrobiol. Hydrogr. 50: 421–463.

Tippett, R., 1978. The effect of the Cruachan pumped storage hydroelectric scheme on the limnology of Loch Awe. University of Glasgow: Unpublished report.

Wedderburn, E. M. & Watson, W., 1909. Current and temperature observations on Loch Ness. Proc. R. Soc. Edinb. 29: 619–647.

West, G., 1910. An epitome of a comparative study of the dominant phanerogamic and higher cryptogamic flora of aquatic habit, in seven lake areas in Scotland. In: Bathymetrical survey of the freshwater lochs of Scotland. Ed. J. Murray & L. Pullar. Edinburgh: Challenger. 156–260.

Whittow, J. B., 1977. Geology and scenery in Scotland. Harmondsworth: Penguin.

4. The phytoplankton

A. E. Bailey-Watts & P. Duncan

Abstract

The seasonal succession of phytoplankton in these five waters is described on the basis of data collected on nine samplings carried out between November 1977 and October 1978. Using chlorophyll a as an index of standing crop, the biomass increases from a winter minimum in each loch (< 0.5 μg l^{-1}) to an annual maximum in summer of ca 1.0 μg l^{-1} in the poorest lochs e.g. Loch Morar, and ca 2.5 μg l^{-1} in the comparatively richer Lochs Awe and Lomond. Considerations of the quality and abundance of the algal floras are restricted to species attaining population densities of at least 10 individuals per litre at some time during the year. Whilst in these soft and nutrient poor waters desmid species are well represented so also are other green algae and blue-green algae; however Cryptophyceae and/or Chrysophyceae constitute the major proportions in terms of numbers of cells recorded at the summer maxima. Fluctuations in the densities of many populations appear to be influenced largely by seasonal changes in water temperature and the extent of thermal stratification. Algal-nutrient relationships are discussed with especial reference to diatoms and silica. Throughout, attempts are made to establish a sequence of lochs illustrating their relative productivity. The results are discussed in relation to general views on oligotrophic lake phytoplankton.

Introduction

This paper describes temporal fluctuations in the abundance of the dominant phytoplankton species of five large, dilute Scottish lochs: Lomond (56°05' N, 4°35' W), Awe (56°20' N, 5°10' W), Ness (57°16' N, 4°30' W), Morar (56°57' N, 5°45' W) and Shiel (56°51' N, 5°30' W). Results are based on a comparative

Monographiae Biologicae, Vol. 44, ed. by P. S. Maitland

study involving nine sampling tours made between November 1977 and October 1978. Together with concurrent physical (Smith *et al.* 1981b), chemical (Bailey-Watts & Duncan 1981a) and other biological studies (Bailey-Watts & Duncan 1981b; Maitland *et al.* 1981b, Smith, *et al.* 1981a, Maitland *et al.* 1981a) this work constitutes a general limnological investigation into the basic nature of these hitherto largely unstudied waters (see also Maitland 1981, and Murray & Pullar 1910 on all the lochs, Jenkin 1930 on Loch Awe, and Slack 1957 and Maitland 1966 on Loch Lomond which is also the site of investigations carried out by Glasgow University from its field station at Rowardennan). Related work has been done on the ecological effects of the pumped-storage hydro-electric schemes on Loch Awe at Cruachan (Tippett 1978; Maitland *et al.* 1981c) and Loch Ness at Foyers (Maitland *et al.* 1981c).

Previously published work on the algal plankton of these waters is limited to the early observations of West & West (1903, 1905, 1909, 1912) and Bachmann (1906, 1907) and the more recent surveys of Brook (1959, 1964, 1965) Boney (1978), and Maulood & Boney (1980). The studies of Tippett (1978) and Maitland *et al.* (1981c) also include algal work, and short-term physiological studies have been made by Maulood *et al.* (1978) on Loch Lomond. The present work thus constitutes the first attempt at a full-year comparative description of plankton fluctuations in these waters.

The Sites

Murray & Pullar (1910) give details of the bathymetry of these lochs. The large size of the water bodies (Table 4.1 – data condensed from Smith *et al.*

Table 4.1   Basic physical features of the five lochs studied (Data from Smith *et al.* 1981b).

Feature	Lomond*	Awe	Ness	Morar	Shiel
Length (km)	19	41	39	19	28
Mean breadth (km)	1.00	0.94	1.45	1.42	0.70
Maximum depth (m)	190	94	230	310	128
Mean depth (m)	78	32	132	87	41
Depth (m) of upper limit of thermocline	8	11	**	17	17
Month of maximum development of thermocline	June	June/July	**	July	June
Lowest deep water temperature (°C)	4.9 (March)	3.4 (Feb.)	5.4 (March)	5.6 (March)	4.7 (Feb.)

* Values refer to the deep northern basin only.   ** No pronounced thermocline develops.

1981b) together with the oceanic climate to which they are subjected, and the low ionic and nutrient status (K_{20} 29-41 μS cm^{-1} and normally < 250 and < 10 μg l^{-1} N and P respectively – Bailey-Watts & Duncan 1981a) are features of major relevance to the composition and growth of their phytoplankton populations. The lochs are of a type whose biological functions are poorly understood in comparison to continental dimictic waters on which most of the current limnological principles have been founded (Ruttner 1963; but cf. Hutchinson 1957, 1967).

Methods

Phytoplankton samples were collected at the times and sites (one open water station per loch), and by the methods described by Bailey-Watts & Duncan (1981a) for their studies on the water chemistry of these lochs. Surface collections thus refer to integrated samples of the top 10 metres of the water column, taken with a weighted polythene tube; six of these were collected at random. Depth samples were taken by closing water bottle – of the Friedinger type – from discrete layers; a single sample was taken at one to three regularly spaced depths in the fully mixed column, and just above and below the thermocline, and in the hypolimnion near the maximum depth during stratification.

Subsamples for cell counts and other microscopic investigations were fixed with acid Lugol's Iodine immediately after collection. Additional material for identification of the larger forms was taken by tow net (180 mesh) and fixed with formalin. Algae were identified with reference to the Fritsch Collection of Algal Illustrations (Lund 1971a) and various taxonomic texts: Bourrelly (1966, 1968, 1970) for all groups; West & West (1904/1923), Brook (1959) and Ruzicka (1977) for desmids; Geitler (1932), Komarek & Ettl (1958) for blue-green algae; Hustedt (1961–1966), for diatoms (and Knudson 1952, 1953a, 1953b for *Tabellaria*); Huber-Pestalozzi (1941) and Bourrelly (1957) for Chrysophyceae; Huber-Pestalozzi (1968) for Cryptophyceae and Dinophyceae, and Lund (1962b) for *Rhodomonas*.

Unfortunately there was little opportunity for examining fresh (un-fixed) cells; failure to name some of the organisms seen was undoubtedly due in part to this. In addition, the recording of many types was based on the detection of only one or a few individuals throughout the study, and the necessary information on cell structure and/or life stages was lacking.

Algae in the iodised samples were allowed to sediment by stages until a 100-fold concentration had been achieved. From these concentrates, cells of

93

Table 4.2 Phytoplankton cell enumeration: probability statements about the detection of species at particular population densities (n ml^{-1}) using the counting procedures described in the text; T relates to the mean count of six 0–10 m tube samples, D the single count from each depth sample.

Probability count		95%		90%		50%	
		T	D	T	D	T	D
Whole chamber		.015	.090	.012	.070	.003	.020
	i.e.	15 l^{-1}	90 l^{-1}	12 l^{-1}	70 l^{-1}	3 l^{-1}	20 l^{-1}
Random grids							
(a) with × 10 objective		2.3	13.8	1.7	10.2	0.5	3.0
(b) with × 25 objective		4.0	24.0	3.0	18.0	0.8	4.8
(c) with × 50 objective		20.0	120.0	15.5	93.0	4.2	25.2

all types were enumerated in a Water Research Association-modified Lund nanoplankton counting chamber (Youngman 1971; Lund 1959a, 1962a). Table 4.2 relates, for these counting methods, the population densities of algae, to the chances of detecting them by at least one 'contact'. Especially rare forms were more likely to be detected in the surface samples, compared to the depth samples, since six times the counting effort was put into estimating numbers in these. The six counts for each species in the tube samples were tested for randomness using X^2 as an index of dispersion. [$s^2/\bar{x} \times (n-1)$; Lund, Kipling & LeCren 1958; Elliot 1977]. The mean counts varied from 0.17 for the rare forms (i.e. a count of one in a total of six samples) to around 120 for dominant forms. The hypothesis of randomness could not be disproved in approximately 80% of cases; of the instances in which non-random distributions were indicated (and these occurrences did not appear to be associated consistently with any particular algal type or season of the year) there were as many indicating 'regular' as 'contagious' distributions. For the great majority of the population estimates presented in this paper therefore, the 95% confidence intervals are reasonably calculated from $\pm$ t $\times$ $\sqrt{\bar{x}/n}$ where t for n = 6 is 2.57. Expressed as a percentage of the maximum and minimum mean counts mentioned above, the intervals range from 10 to 257.

Aliquots of water for chlorophyll *a* estimations on the surface and outflow samples were filtered through Whatman glass-fibre discs according to the method outlined by Talling (1965, see also Tolstoy, 1977); 3 l of these algal-poor waters were usually necessary; two estimates were carried out on water bulked from the six tube samples, and a third on one of the pair of outflow samples. The spectrophotometric equation of Talling & Driver (1963) was used to relate concentrations of chlorophyll *a* to the absorbances of methanolic

94

extracts of algae at 665 nm. No corrections were made for pheophytin interference.

Results

Firstly in this section, the general character of the phytoplankton populations is illustrated with seasonal information on crop abundance and species composition. The possible influences of environmental factors on phyto- plankton population structure are considered next; trends in algal distribution with time and depth are related to physical factors, particularly thermal layering. Finally, the influence of water chemistry is considered in terms of both the general ionic composition and nutrient levels. Throughout, an attempt is made at a broad comparison of the algal populations of the five lochs, with fine details of individual species differences being omitted.

Seasonal changes in algal crops: chlorophyll a *and total numbers of indi- viduals*

Chlorophyll *a* concentration was used as an index of total algal biomass and Fig. 4.1 shows the seasonal trends found in the lochs. Each pattern consti- tutes a monacmic type (Ostenfeld 1913 in Hutchinson 1967) with basically a single peak, occurring in summer. The levels of chlorophyll recorded were very low, the highest being only 3.2 μg l^{-1} in Loch Awe; this occurred in June and preceded the August maxima in all the other lochs. The gradual increase in chlorophyll levels towards the maximum commenced earliest in Loch Awe, with levels in March double those of nearly all the other waters. If the lochs are placed in increasing order of annual algal biomass – on the basis of chlorophyll *a* maxima and yearly mean levels – the first is Loch Morar, followed by Lochs Shiel, Ness, Lomond and Awe.

Estimates of the total number of cells present at the times of the chlorophyll *a* maxima further confirm the relative richness of the lochs e.g. Loch Awe in June is very much richer than Loch Morar in August (Table 4.3): however only by excluding the counts of small chrysoflagellates does the rest of the sequence reflect that of the chlorophyll levels (Fig. 4.2). Even with this modification, Loch Shiel appears relatively richer than on the basis of the pigment data. Total algal cell numbers can give only a very broad indication of the different amounts of matter present however, since the biomass of many small cells equals that of a few large organisms. In addition, algae within some

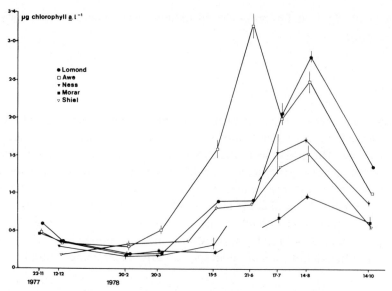

Fig. 4.1 Fluctuations in the concentration of chlorophyll *a*. Values refer to the integrated 0-10 m column at the open water site of each loch. Vertical lines indicate the difference between the pair of analyses carried out on each occasion, although in some cases this is less than the dimension of the symbols used. Dates on the x axis refer to the Wednesday of each week-long sampling tour.

Table 4.3 Total numbers of algae present in the five lochs at the times of their chlorophyll *a* maxima shown in Fig. 4.1

Loch	Total algal numbers* (1000's l⁻¹)
Lomond	528
Awe (June)	1691
(August)	470
Ness	568
Morar	123
Shiel	1005

* Numbers of 'entities' i.e. colonies (except for colonial pennate diatoms which have been included as total cells), filaments, unicells etc.

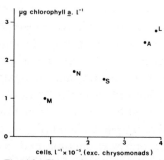

Fig. 4.2 The relationship between chlorophyll *a* concentration and total number of phytoplankton individuals (excluding chrysomonads) at the time of the August pigment maxima in each loch (L = Lomond; A = Awe; N = Ness; M = Morar; S = Shiel).

groups e.g. Chlorococcales tend to be richer in chlorophyll per unit cell volume or mass than others e.g. diatoms (Bursche 1961) whilst variation in environmental light intensity and nutrient supply may effect pigment changes within a single individual.

96

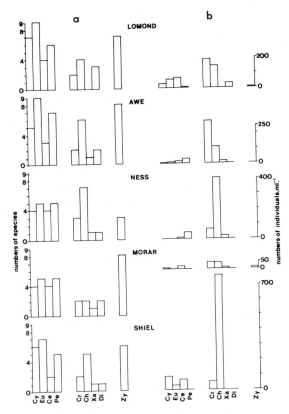

Fig. 4.3 The distribution, within nine major algal Classes, of species attaining population densities of at least 10 individuals l⁻¹ sometime during the study by (a) numbers of species recorded throughout the study and (b) numbers of individuals at the time of the August pigment maxima. (Cy = Cyanophyceae; Eu = Euchlorophyceae; Ce =centric diatoms; Pe = pennate diatoms; Cr = Cryptophyceae; Ch = Chrysophyceae; Xa = Xanthophyceae; Di = Dinophyceae; Zy = Zygophyceae (desmids)).

Qualitative differences

Data in Table 4.3 and Fig. 4.2 concern only those species attaining densities of at least 10 individuals l⁻¹. The same population density minimum was used in constructing Fig. 4.3, which indicates the qualitative composition of the five floras. Fig. 4.3a shows the total number of species in each algal group encountered in the year's sampling, and Fig. 4.3b compares the total number of cells in each algal group during the August chlorophyll maxima referred to previously. This approach enables a straight comparison to be made between

97

lochs in terms of either a wide range of forms including relative rarities, as in Fig. 4.3a, or a more restricted assemblage of the most abundant species which control the general form of the bar charts in Fig. 4.3b.

The two relatively 'richer' waters (Lochs Lomond and Awe) exhibit the greatest total numbers of species – 42 and 43 respectively, compared with 33 in Ness and Morar and 35 in Shiel (Fig. 4.3a). Each loch differs from the rest in the relative proportion of species by groups; no one group of algae is consistently better or less well represented in the 'richer' or 'poorer' lochs. A notable feature is that whilst desmids are generally well represented in these soft waters, both the green and blue-green groups are also important. None of the three groups however is very evident in Fig. 4.3b which illustrates the biomass dominance by other forms, e.g. cryptomonads in Lochs Lomond and Awe and chrysoflagellates in Lochs Ness and Shiel.

Species population succession

Fig. 4.4-4.6 show the seasonal abundances of fifteen forms recorded in all five lochs, thus permitting a full comparison between lochs. Data on eight of these have been grouped at the generic level: the blue-green alga *Oscillatoria* (including species of the *agardhii-rubescens* type, Fig. 4.4a); two green algae *Elakatothrix* (including *E. gelatinosa* Willé but with some specimens resembling *Quadrigula closterioides* (Bohlin) Printz, Fig. 4.4b) and *Oocystis* (including *O. crassa* Wittr., *O. gloeocystiformis* Borge and *O. rhomboidea* Fott., Fig. 4.4c); two centric diatoms *Melosira* (including *M. islandica* O. Müll., *M. italica* subsp. *subarctica* O. Müll., and *M. granulata* (Ehr.) Ralfs, Fig. 4.4d) and *Rhizosolenia* (*R. longiseta* Zach., and *R. eriensis* H. L. Smith, Fig. 4.4e), two cryptomonads *Cryptomonas* (*C. erosa* Ehrenb., and *C. ovata* Ehrenb., Fig. 4.5a) and *Rhodomonas* (*R. minuta* var *minuta* Skuja and var *nannoplanctica* Skuja – see Lund 1962b – Fig. 4.5b); and finally, the chrysoflagellate *Mallomonas* (near *M. caudata* Iwanoff and *M. acaroides* Perty, Fig. 4.5c).

Fig. 4.4f shows the abundances of small unicellular centric diatoms which, in the absence of sufficient material for definitive identifications to be made, were plotted as a group; however *Cyclotella comta* (Ehr.) Kütz., *C. glomerata* Bachmann and *Stephanodiscus astraea* (Ehr.) Grün are major constituents. The remaining graphs are of six other species common to all the lochs: three colonial pennate diatoms (*Asterionella formosa* Hass., *Tabellaria fenestrata* (Lyngb.) Kütz including var *asterionelloides* (Roth) Kütz (but cf. Knudson, 1952, 1953a) and *Tabellaria flocculosa* (Roth) Kütz (Fig. 4.4g-i), a naked

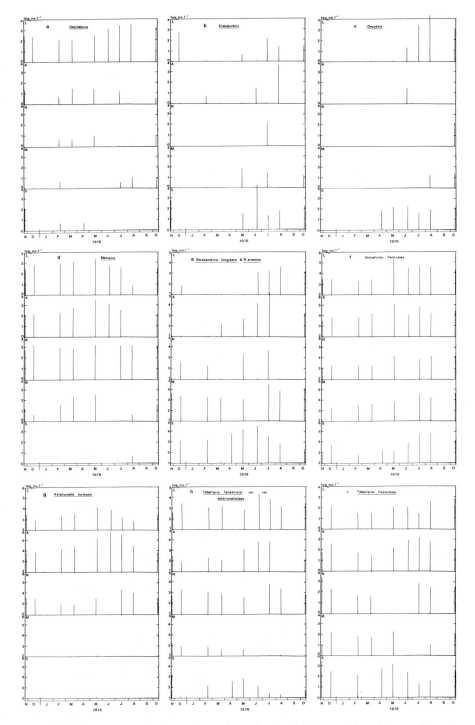

Fig. 4.4a-i Fluctuations in the numbers of *Oscillatoria* trichomes, *Elakatothrix* and *Oocystis* colonies, *Melosira* threads, *Rhizosolenia* cells, unicellular Centrales and cells of *Asterionella* and *Tabellaria*. Values refer to the 0-10 m column at the open water site in each loch; (L = Lomond; A = Awe; N = Ness; M = Morar; S = Shiel).

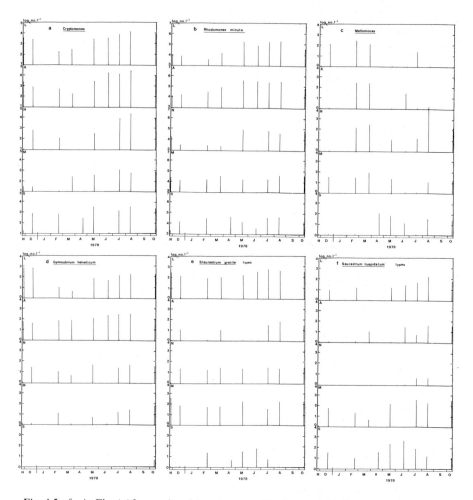

Fig. 4.5a-f As Fig. 4.4 for species of *Cryptomonas, Rhodomonas, Mallomonas, Gymnodinium* and *Staurastrum.*

dinoflagellate (*Gymnodinium helveticum* Penard but very near var *achroum* Skuja – Fig. 4.5d) and two groups of desmids, both *Staurastrum* – (short- and long-armed varieties of *S. gracile* Ralfs – Fig. 4.5e – and *S. cuspidatum* Bréb types – Fig. 4.5f).

Some of these algae are very sparse in certain locations. *Melosira* and *Gymnodinium* are rare in Loch Shiel, *Oscillatoria* is well represented only in Lochs Lomond and Awe, and the two green algae, *Elakatothrix* and *Oocystis,* show a constant but fluctuating background presence to the dominant forms rather than being especially important themselves.

100

Whilst the phytoplankton of the lochs is best characterised by records of the above forms, five other types merit special mention: *Ankistrodesmus* (including a number of forms of *A. falcatus* (Corda) Ralfs) *Botryococcus braunii* Kütz, *Coelosphaerium naegelianum* Unger, nanochrysoflagellates (including *Chromulina* and *Ochromonas* types) and a *Bumilleriopsis*-like xanthophyte, were extremely important in some lochs if only for short periods, but were not recorded in others (Fig. 4.6). The two green algae were 'absent' from Loch Ness only, and the blue-green alga was evident in only the two 'richest' waters – Lochs Awe and Lomond. Some species from Loch Lomond are illustrated in Plate 4.1. Other forms were recorded but these are relatively unimportant in terms of biomass.

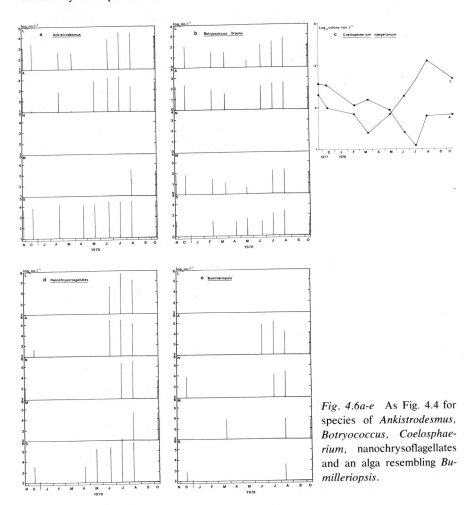

Fig. 4.6a-e As Fig. 4.4 for species of *Ankistrodesmus, Botryococcus, Coelosphaerium,* nanochrysoflagellates and an alga resembling *Bumilleriopsis.*

101

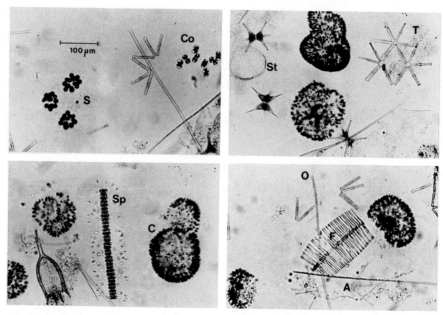

Plate 4.1 Some tow-netted planktonic algae from Loch Lomond: genera represented include *Coelosphaerium* and *Oscillatoria* (bluegreen algae C,O), *Tabellaria, Asterionella* and *Fragilaria* (pennate diatoms T, A, F), *Sphaerocystis* (colonial green alga S), and *Cosmocladium, Spondylosium* and *Staurastrum* (desmids Co, Sp and St). (Photos: A. E. Bailey-Watts.)

The influence of physical and chemical factors on fluctuations in the phytoplankton

Physical factors. Changes in the levels of chlorophyll *a* and numbers of the majority of the algae mentioned above show similar patterns. Concentrations of the pigment, and numbers of *Oscillatoria, Cryptomonas, Rhodomonas*, desmids, certain green algae and many diatoms, are all lowest in winter and rise to maxima during the summer. Temperature data, taken from Smith *et al.* (1981b) are presented in Fig. 4.7 and show that the period of algal increase coincides with the development of thermal stratification. This suggests that associated conditions – presumably related to increases in temperature and the decreasing extents of vertical mixing, in turn affecting the light regime (see eg. Talling 1971) – become more favourable for algae during this period; in effect, the lochs become temporarily shallower and warmer. At the height of stratification (July-August) epilimnion depths in the lochs vary from 8 m in Loch Lomond to 17 m in Lochs Morar and Shiel (Table 4.1).

Responses to this major physical change follow one of three basic patterns

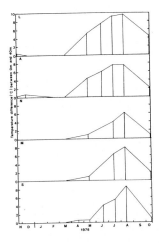

Fig. 4.7 Fluctuations in the extent of thermal stratification as indicated by surface and 40 m values of temperature in the five lochs (L = Lomond; A = Awe; N = Ness; M = Morar; S = Shiel).

according to species and are indicated by the plots of population change in the top 10 m of the water column. The first is exhibited by forms that maintain a net increase during this period. These include the blue-green algae *Oscillatoria* (Fig. 4.4a) and *Coelosphaerium* (Fig. 4.6c – in Loch Lomond but not Loch Awe), green algae *Oocystis* (Fig. 4.4c), *Ankistrodesmus* (Fig. 4.6a), and *Botryococcus* (Fig. 4.6b), the two *Staurastrum* species (Fig. 4.5e and f) and *Cryptomonas* (Fig. 4.5a). A second type of response is one in which numbers fluctuate around the highest levels for the year during this period. but show no consistent increase or decrease. This is evident in the plots of *Rhizosolenia* (Fig. 4.4e), small unicellular centric diatoms (Fig. 4.4f), *Tabellaria flocculosa* (Fig. 4.4i), *Rhodomonas* (Fig. 4.5b), *Gymnodinium* (Fig. 4.5d), small chrysoflagellates (Fig. 4.6d) and, in Loch Awe, the xanthophyte (Fig. 4.6e).

The third response, by which cell numbers appear to decrease in the surface waters as the water column stabilises, is suggested most markedly by *Melosira* (Fig. 4.4d) an especially heavy phytoplankter (Lund 1959b 1964) whose distribution is known to be controlled at least partly by thermal patterns (Lund 1954, 1955, 1971b). *Tabellaria fenestrata* appeared also to decrease at the surface (Fig. 4.4h). That these and other diatoms sink to considerable depths in the water column is indicated in Fig. 4.8a-d. However, the scarcity of cells at deeper points (depth data for the xanthophyte and *Rhodomonas* are included in Fig. 4.9 to illustrate this) does not necessarily exclude the possibility of sinking; the cells of 'soft' forms such as flagellates are likely to fragment and thus disappear from view more rapidly than e.g. diatom cells or at least their frustules (see Discussion). Nevertheless, the distributions of these small forms commonly show greater values in the surface waters whilst the reverse is the case with threads of *Melosira*.

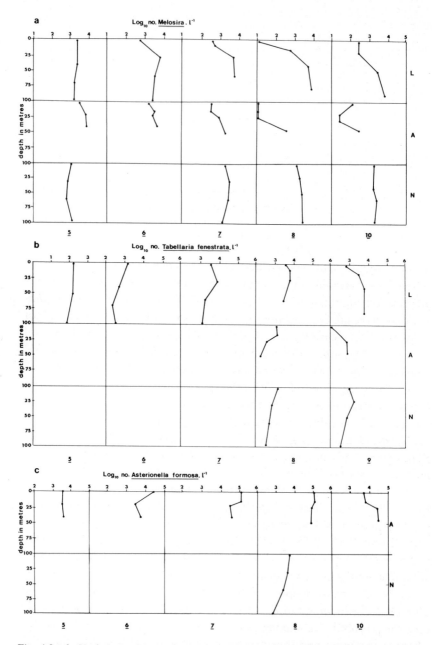

Fig. 4.8a-d Variations with depth in Lomond (L), Awe (A), Ness (N), Morar (M) and Shiel (S) in the abundance of *Melosira* threads, *Tabellaria fenestrata* and *Asterionella formosa* colonies, and unicellular centric diatoms. Numbers underlined refer to the sampling month. (See next page for 4.8d.)

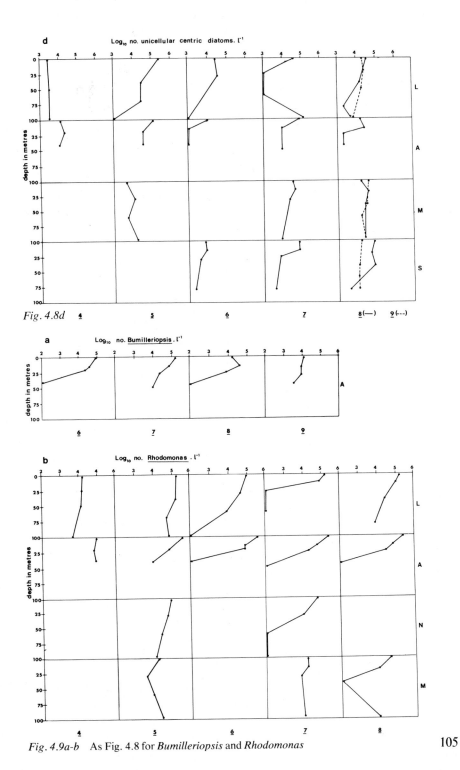

Fig. 4.8d

Fig. 4.9a-b   As Fig. 4.8 for *Bumilleriopsis* and *Rhodomonas*

105

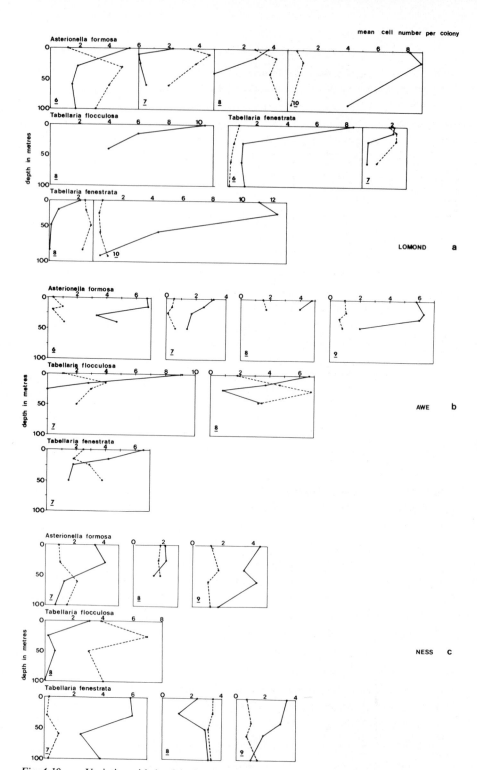

Fig. 4.10a-c Variation with depth in the number of cells (——live, –––dead;) per colony of three pennate diatoms in Lochs Lomond, Awe and Ness. Numbers underlined refer to the sampling month.

No such marked increases in number with depth were found with *Asterionella* or *Tabellaria fenestrata* colonies in spite of decreases in their numbers recorded in the surface water of at least some of the lochs. The condition of the cells and colonies of these species, however, did vary with depth. Fig. 4.10 shows the consistent decreases with depth in the numbers of live cells per colony, as well as changes in the incidence of dead cells.

Chemical factors. Generally richer phytoplankton assemblages are found in Lochs Lomond and Awe with similar totals of species (42 and 43 respectively – Fig. 4.3a) and numbers of individuals at the August maxima (*ca.* $400 \times 10^3 \times l^{-1}$ – Fig. 4.2). These lochs are the richest of the five in terms of alkalinity, and pH (Table 4.4) and nutrients (Bailey-Watts & Duncan 1981a).

Table 4.4 Some chemical features of the five lochs listed in order of increasing algal biomass. Values are means for the November 1977-October 1978 period

Loch	Conductivity ($\mu S\ cm^{-1}$ at 20 °C)	Alkalinity (mg $CaCO_3\ l^{-1}$)	pH	Ca	Mg	Na	K	Total Cations	Ca + Mg ——— Na + K
				←—————(mg l^{-1})—————→					
Morar	35	3.27	6.63	1.23	0.88	5.34	0.34	7.79	0.55
Shiel*	29	2.10	6.14	0.98	0.75	4.24	0.25	6.22	0.58
Ness	30	4.48	6.70	1.99	0.81	3.84	0.27	6.91	0.96
Lomond	34	6.37	6.78	3.00	1.00	3.45	0.33	7.78	1.47
Awe	41	8.97	6.90	4.01	0.99	4.47	0.27	9.74	1.41

* The position in this series of Shiel is not well-established; it is placed here in accordance with chlorophyll *a* levels (see Text).

Loch Morar and the north basin of Loch Lomond both lie in the middle of the range of total cationic strength and conductivity (Fig. 3.2 in Bailey-Watts & Duncan 1981a). However, Loch Morar is the poorest of all the lochs in algal numbers, although in terms of species it is similar to Loch Ness (each 33 species in Fig. 4.3a). Loch Shiel is the most dilute loch, yet it supports a comparatively abundant crop of phytoplankton in which blue-green algae (dominated by *Merismopedia* spp.) are more evident than in any of the other lochs during the August maxima (Fig. 4.3).

The mean concentrations of inorganic nitrogen are higher in Loch Lomond (0.2 mg nitrate plus nitrite N l^{-1}) than the next richest water. Loch Awe (0.11 mg l^{-1}), although the total numbers of algae recorded in each loch are fairly similar with Loch Awe exhibiting the higher chlorophyll levels. Dissolved inorganic phosphorus levels in all lochs including these two richer waters were consistently below the detection level (10 μg P l^{-1}) of the analytical methods used (Bailey-Watts & Duncan 1981a). Further ecological complexities are indicated by the data on silica concentration and diatom abundance. A simple relationship between silica and diatoms is not always found (Bailey-Watts 1976a) but many studies (see e.g. Lund 1950a, 1950b, 1955, Jorgensen 1957, Hutchinson 1967, Bailey-Watts 1976b) have indicated strong causal interactions between the two. However, in the present work, only in Lochs Lomond and Awe do the major net increases in total diatom numbers coincide with the main decrease in dissolved silica (March to May – Bailey-Watts & Duncan 1981a). Depletions of dissolved silica occur over the same period in the other three lochs, but the diatoms increase later than May.

Further inconsistencies are revealed when a balance is attempted between changes observed in dissolved silica and those expected in terms of diatom opal and thus approximate cell numbers. March to May decreases in Si in the surface waters of these lochs range from 500 μg l^{-1} in Loch Shiel to 770 μg l^{-1} in Loch Lomond. If such changes can be attributed solely to diatom utilisation, then, using the silica content of *Asterionella* as a guide (150 μg Si O$_2$ 10^{-6} cells – see Lund 1965) population increases equivalent to 6.6-10.2 $\times$ 10^6 cells l^{-1} could be expected. In the absence of massive losses by zooplankton grazing, sinking out of the water column and passage down the outflows, these figures give an approximate guide to the expected increase in the diatom population. However, as Fig. 4.11 shows, the largest diatom increase which occurred in Loch Awe from May to June, amounted to only 0.65 $\times$ 10^6 cells l^{-1}. Furthermore, as the majority of the cells contributing to this increase were small centric diatoms which contain less silica than *Asterionella*, they would actu-

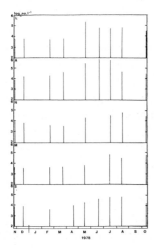

Fig. 4.11 Total diatom numbers in Loch Lomond (L), Awe (A), Ness (N), Morar (M) and Shiel (S). Values refer to numbers of cells (except for thread-like diatoms which are included as threads), in integrated 0-10 m tube samples.

108

ally account for an even smaller proportion of the observed losses of the dissolved element.

Discussion

This section considers firstly some factors influencing the status of the five lochs in terms of algal biomass. Secondly, the more general features of the data and some of their apparent inconsistencies are discussed. Finally the whole is considered in relation to current views on the character of oligotrophic waters in general.

A large number of factors influence the qualitative and quantitative aspects of phytoplankton development in fresh waters (Lund 1965). This may explain why the attempts to place these five lochs in sequences according to the relative importance of factors thought to influence phytoplankton growth, have resulted in a different loch sequence for each chosen parameter. A multivariate analytical approach is thus necessary to assess the correct weightings for the various interactions involved. For example, while the generally dense algal crops of Lochs Lomond and Awe are associated with the greater ionic strengths of these waters, the related slightly richer nutrient concentrations are likely to be the more important factors. It was found that Loch Morar is rather similar to Loch Lomond in terms of conductivity and total cations. Because, however, the major components of the chemistry of Loch Morar water (sodium and chloride – Bailey-Watts & Duncan 1981a) are likely to be largely oceanic rather than terrestrial in origin, the concomitant introduction of plant nutrients does not take place and thus algal growth is limited.

If the number of algal individuals observed is a broad index of photosynthetic productivity, then even the dark and deep column of Loch Ness in the series is one of the more suitable for algal growth. While lower algal productivity in Lochs Ness and Morar are in part at least associated with the immense depths of these basins, the dark coloration of Loch Ness water must impose an additional stringency. None of the phytoplankton crops observed was dense enough for the limiting affects of 'self-shading' (Talling 1960, 1970) to be important, but in both Lochs Ness and Awe, water clarity was influenced largely by dissolved organic ('humic') material, and light attenuation was marked in these two lochs (Bailey-Watts & Duncan 1981a). It is possible that the high background colour of Loch Awe water to some extent negates the theoretically greater algal production potential of this loch as compared with the much deeper Loch Lomond. However, as the vertical profiles of temperature indicate, (Smith, *et al.* 1981b) the epilimnion of Loch Lomond was considerably shallower than that of Loch Awe. Gibson, *et al.* (1980) illustrate light limitation of algal growth in Lough Erne (N. Ireland) engendered by a deep mixed column of peaty water; these authors found spatial differences in algal quality and abundance that could be related to the variation in mixed depth in different parts of the lough.

A comparison of the lochs on the basis of algal biomass per unit area of water surface was not attempted since a programme of more intensive depth sampling would have been required. However, it is likely that production beneath unit area in the deeper Lochs Ness and Morar is considerably less than in e.g. Loch Awe. Features of the five basins are further compared by Maitland *et al.* (1981d) in relation to the various biota recorded.

Loch Shiel has not been mentioned so far in this discussion. Albeit only in a general way, these studies indicate that of the five waters, Lochs Lomond and Awe form a similar pair at the 'rich' end of our scales while Lochs Ness and Morar constitute the 'poor' end. Loch Shiel however exhibits no such consistency. Chemically it is the most dilute being low in all ions and thus conductivity. However, even with low nutrients it supports as great a number of species as Lochs Morar and Ness, and in summer at least, produces some of the densest populations af algae, particularly chrysoflagellates. On the other hand it is, but for Loch Awe, the shallowest of the basins studied, a feature which may offset the apparent relative lack of nutrients. The incident light regime at the Loch Shiel sampling station is also perhaps the least favourable for algal growth. An especially high horizon due to the topography of surrounding land, causes a complete cutout of *direct* sunlight near the shortest day of the year in December at latitude 57° N and effects a 2-h shortening of the days even in mid-summer (Smith *et al.* 1981b).

A general feature of many data is the marked seasonal trends, in spite of the choice of a single sampling point and the relative infrequency of sampling. That this schedule has resulted in a loss of detailed seasonal data is undoubted. Maulood *et al.* (1978) in one of the relatively few really intensive studies in northern temperate waters, (cf. tropical works e.g. Ganf & Horne 1975) showed that in Loch Lomond, changes in algal numbers, nutrients and physical variables could be detected over time intervals as short as one hour. These observations were made at a point where the north and south basins of Loch Lomond meet i.e. to the south of the station used in this investigation.

Although these studies and the work of Tippett (1978) on Loch Awe, and George & Jones (in preparation) on each of these five waters, show that spatial variations in almost all variables do occur (see also e.g. Small 1963, George & Edwards 1976, George & Heaney 1978), it is unlikely that any major features of the plankton of these waters have been missed by our approach.

Tippett (1978) showed the southern basin of Loch Awe to be more productive than its more northerly basin and northernmost arm. This was evidenced by differences in maximal algal numbers, species diversity and the observed depletion of nutrients. Maxima of total algal numbers in Loch Awe during the present study approximated to 1.7×10^6 l^{-1} in June and 4.7×10^5 l^{-1} in August; the sampling point corresponded to Tippett's 'No. 1' station, where in 1975 the maximal algal number was similar – 8.5×10^5 l^{-1}. Similarities in algal species composition are also evident between these results and those of Maulood & Boney (1980) on Loch Lomond ('net' plankton) and of Tippett (1978) on Loch Awe and his on-going observations on Loch Lomond (Dr R. Tippett – personal communication). Such consistency between the findings of entirely independent research programmes is gratifying but perhaps not so surprising in view of the size of the lochs in question. These waters are extremely well 'buffered' in the physical sense. Gross changes in chemical character, and at least the more direct biological consequences of it, are unlikely to take place over very short times spans and year-to-year differences are liable to be relatively minor (but cf. differences in *Tabellaria* abundance in two consecutive years in Loch Lomond – Maulood & Boney 1980) compared to the gross changes that may occur quite rapidly in shallow lochs (Bailey-Watts 1978). The similar findings of the different studies is particularly notable, however, in view of the considerable difficulties inherent in estimating micro-algal numbers in these lochs. Not only are phytoplankters very sparse here, but the algal mass commonly comprises only a mere fraction of the total suspended matter present. This is indicated by total organic carbon estimates of around 2 mg C l^{-1} in even the clearer waters (Lochs Lomond, Morar and Shiel – Bailey-Watts & Duncan 1981a).

'Interference' due to non-algal suspended organic matter, and the sampling intervals employed, may have affected one of the main inconsistencies of our data viz. that concerning the poor agreement between the observed diatom maxima and the concomitant dissolved silica depletions. One possible explanation is that the real maxima occurred during one of the long time intervals between samplings. That algal numbers changed quite rapidly in summer – the season for which the diatom-silica balances have been attempted – is evident from our work and that of Tippett (1978). A further possibility, however, is that silica was being utilised or taken up by agencies other than planktonic diatoms. Even in such large bodies of water, littoral algae, including diatoms, do occur (e.g. varieties of *Tabellaria flocculosa* in Loch Ness, Knudson 1953a, and other species evident from our lists and those of

111

Tippett 1978). Other silica-containing algae e.g. *Mallomonas* (Chrysophyceae) are abundant, but it is not thought that these can account for the major additional utilisation.

In attempting to identify the cause of the apparently large silica losses, a process rapidly affecting the whole water mass should be considered. A chemical mechanism involving a linkage of silica to organic substances is possible (see e.g. Mortimer 1942, Tessenow 1964) but this, as well as many other aspects of silica chemistry, seems to be poorly understood.

This discussion concludes with a review of the current general concepts concerning certain aspects of oligotrophic lake phytoplankton. None of these lochs could be termed nutrient-rich (see Vollenweider 1968) – indeed, it is the general sparseness of phytoplankton that fits with the concept of morphometric and edaphic oligotrophy (Rawson 1955, 1956). In general the population densities of shallower, more nutrient-rich lakes may exceed those recorded during this study by factors of 10^2 or 10^3 (see e.g. Gibson, Wood, Dickson & Jewson 1971 for Lough Neagh, N. I.; Bailey-Watts 1978, for Loch Leven, Scotland; Lund 1949, 1954, 1961b, 1962b, 1971c for Blelham Tarn, Esthwaite Water and Lake Windermere, England; and e.g. Ridley 1970, Youngman 1975 and Collingwood 1977 for English water supply reservoirs). Also, the appearance of only one major crop maximum in the year may be expected of the waters studied here (Hutchinson 1967). When the phytoplankton species composition is considered however, no such striking dissimilarities from even eutrophic lake floras are evident. Numerous freshwater studies (but see especially Moss 1972, 1973a,b,c) show that no lake type is the preserve of any particular algal species or group to the exclusion of others. Chance dispersal of organisms between lakes is also likely to mask the identity of the truly native members (see e.g. Knudson 1954, 1955 in relation to *Tabellaria* species).

Rawson (1956) questions the very existence of species indicating oligotrophic conditions, and both he and Brook (1965) discuss the relative merits of dominant and rare forms as indicators. An argument against the use of the dominant forms is that these are often distributed widely among other waters of many different types. In this connection the overall similarity between the planktonic floras of these five lochs is of interest, although to confirm this, more attention needs to be paid to possible differences between the lochs in species and varieties, especially within taxonomically difficult genera e.g. *Melosira, Tabellaria, Mallomonas, Cryptomonas*.

In view of the emphasis on desmids as indicators of soft-water conditions and oligotrophy (see e.g. Nygaard 1949, Brook 1964, 1965, West & West 1903, 1905, 1909, Pearsall 1932, but cf. Moss 1972, 1973a,b,c) the rather scant

112

attention given to this group in the results of the present study may appear parodoxical. It should be remembered, however, that the early workers in particular drew up species lists on the basis of algal collections made with a tow net. The present study hints at the great variety of species including desmids that might have been identified if algae at much lower densities had been fully investigated, however these forms have been largely ignored by concentrating on species attaining a minimum population density of 10 individuals l^{-1}. Phytoplankton ecologists are becoming increasingly aware of oligotrophic water forms from otherwise eutrophic indicator groups (such as blue-green algae) and conversely, eutrophic indicators from soft-water groups (such as desmids). However, collection of basic autecological knowledge of all species will be necessary before a true understanding of freshwater algal distribution is gained. In view of our ignorance in these matters, it is perhaps not surprising that only a very rudimentary ordering of these five lochs on the basis of their phytoplankton has been possible.

One school of thought suggests that more species of algae are found in oligotrophic than eutropic lake phytoplankton. This may be the case when highly eutrophic or heavily polluted waters are included; a gradual reduction of species is observed during the development of algal maxima in eutrophic lakes (Bailey-Watts 1973). However, on comparing the algal lists published by e.g. Lund (1961a for Malham Tarn, 1964 for Blelham Tarn and Esthwaite Water), Reynolds (1973 for Crose Mere, Shropshire), Bailey-Watts (1974 for Loch Leven) i.e. for much richer waters, with the number of species mentioned in this coverage of the five oligotrophic lochs, the suggestion that the latter are more species rich is unfounded. This holds in spite of lack of identification of all species encountered here. Results of the present study also lend little support to the observations of other workers that oligotrophic phytoplankton tends to be dominated by nano, rather than net forms (see e.g. Lund 1961b, Pavoni 1963, Kristiansen 1971). However, such views are largely subjective, being fraught with the vagaries of assessing which species are 'net' forms and which are not. Contrastingly, an objective analysis of phytoplankton succession in terms of size of cell, filament, colony etc., is likely to yield a wealth of information on the influences of water movements and nutrient levels on algal growth, and the contributions by zooplankton grazing and sedimentation to algal losses (Bailey-Watts & Kirika 1981). In this connection the abundance of small crypto-flagellates observed over much of the year, and chrysomonads in summer in these waters, suggests that grazing pressure from zooplankton is not important.

Acknowledgements

The programme described forms part of a study carried out under contract to the North of Scotland Hydro-Electric Board and the Nature Conservancy Council. We are particularly grateful to Mr A. J. Rosie for help with all the field sampling and Dr C. E. Gibson for valuable comments on this paper. To them and other colleagues – especially Dr P. S. Maitland, Mr I. R. Smith, Ms B. D. Smith and Mr A. A. Lyle – go thanks for many constructive and fruitful discussions throughout the course of the work. Mr A. Kirika is gratefully acknowledged for much help with data analyses and for carrying out most of the chlorophyll determinations. We thank Dr R. Tippett for allowing us to consult his lists of Loch Lomond phytoplankton. To Ms G. M. Dennis and Mr M. J. Carr go especial thanks for drawing the figures and obtaining copies of literature respectively. Finally we thank Mrs M. S. Wilson for typing the manuscript.

References

Bachmann, H., 1906. Le plankton des lacs écossais. Archs. Sci. phys. nat. (Genève) 20: 359.

Bachmann, H., 1907. Vergleichende studien über das Phytoplankton von Seen Schottlands und der Schweiz. Arch. Hydrobiol. Planktonk. 3: 1–91.

Bailey-Watts, A. E., 1973. Observations on the phytoplankton of Loch Leven, Kinross, Scotland. Ph.D. Thesis. University of London. 337 pp.

Bailey-Watts, A. E., 1974. The algal plankton of Loch Leven, Kinross. Proc. R. Soc. Edinb. B. 74: 135–156.

Bailey-Watts, A. E., 1976a. Planktonic diatoms and some diatom-silica relations in a shallow eutrophic Scottish loch. Freshwat. Biol. 6: 69–80.

Bailey-Watts, A. E., 1976b. Planktonic diatoms and silica in Loch Leven, Kinross, Scotland: a one month silica budget. Freshwat. Biol. 6: 203–213.

Bailey-Watts, A. E., 1978. A nine-year study of the phytoplankton of the eutrophic and non-stratifying Loch Leven (Kinross, Scotland). J. Ecol. 66: 741–771.

Bailey-Watts, A. E. & Duncan, P., 1981a. The ecology of Scotland's largest lochs: Lomond, Awe, Ness, Morar and Shiel. Ed. P. S. Maitland. Chemical characteristation – a one-year comparative study. Chap. 3. This volume.

Bailey-Watts, A. E. & Duncan, P., 1981b. The ecology of Scotland's largest lochs: Lomond, Awe, Ness, Morar and Shiel. Ed. P. S. Maitland. A review of macrophyte studies. Chap. 5. This volume.

Bailey-Watts, A. E. & Kirika, A., 1981. Assessment of size variation in Loch Leven phytoplankton: Methodology and some of its uses in the study of factors influencing size. J. Plank. Res.

Boney, A. D., 1978. Microscopic plant life in Loch Lomond. Glasg. Nat. 19: 391–402.

Bourrelly, P., 1957. Recherches sur les Chrysophycées. Revue algol. Memoire 1.

Bourrelly, P., 1966. Les Algues d'Eau Douce. I. Les Algues Vertes. Paris: Boubée et Cie.

Bourrelly, P., 1968. Les Algues d'Eau Douce. II. Les Algues Jaunes et Brunes, Chrysophycées, Phéophycées, Xanthophycées et Diatomées. Paris: Boubée et Cie.

Bourrelly, P., 1970. Les Algues d'Eau Douce III. Les Algues Bleues et Rouges, Les Eugleniens, Peridiniens et Cryptomonadines. Paris: Boubée et Cie.

Brook, A. J., 1959. *Staurastrum paradoxum* and *S. gracile* in the British freshwater plankton and a revision of the S. anatinum group of radiate desmids. Trans. R. Soc. Edinb. 63: 589–628.

Brook, A. J., 1964. The phytoplankton of the Scottish freshwater lochs. In: The vegetation of Scotland (ed. J. H. Burnett,) Edinburgh: Oliver & Boyd.

Brook, A. J., 1965. Planktonic algae as indicators of lake types with special reference to the Desmidiaceae. Limnol. Oceanogr. 10: 403–411.

Bursche, E. M., 1961. Anderungen in Chlorophyllgehalt und in Zellvolumen bei Planktonalgen, hervorgerufen durch unterschiedliche Lebensbedingungen. Int. Rev. ges. Hydrobiol. Hydrogr. 46: 610–652.

Collingwood, R. W., 1977. A survey of eutrophication in Britain and its effects on water supplies. Tech. Rep. Wat. Res. Centre, TR40.

Elliott, J. M., 1977. Some methods for the statistical analysis of samples of benthic invertebrates. Freshwater Biological Association Scientific Publication. No. 25.

Ganf, G. G. & Horne, A. J., 1975. Diurnal stratification, photosynthesis and nitrogen-fixation in a shallow, equatorial lake (Lake George, Uganda). Freshwat. Biol. 5: 13–39.

Geitler, L., 1932. Cyanophyceae In Rabenhorst, L., Kryptogamenflora von Deutschland, Österreich und der Schweiz, 14: Leipzig.

George, D. G. & Edwards, R. W., 1976. The effect of wind on the distribution of chlorophyll *a* and crustacean plankton in a shallow eutrophic reservoir. J. appl. Ecol. 13: 667–690.

George, D. G. & Heaney, S. I., 1978. Factors influencing the spatial distribution of phytoplankton in a small productive lake. J. Ecol. 66: 133–155.

George, D. G. & Jones, D. H., 1981. Spatial studies of the zooplankton of Scotland's largest lochs: Lomond, Awe, Ness, Morar and Shiel. In preparation.

Gibson, C. E., Foy, R. H. & Fitzsimons, A. G., 1980. A limnological reconnaissance of the Lough Erne System, Ireland. Int. Revue ges. Hydrobiol. Hydrogr. 65: 49–84.

Gibson, C. E., Wood, R. B. Dickson, E. L. & Jewson, D. H., 1971. The succession of phytoplankton in Lough Neagh 1968–70. Mitt. int. Verein. theor. angew. Limnol. 19: 146–60.

Huber-Pestalozzi, G., 1941. Das Phytoplankton des Susswassers 2:1 Chrysophyceen, Farblose Flagellaten, Heterokonten. In: Thienemann A., Die Binnengewasser XVI. Stuttgart.

Huber-Pestalozzi, G., 1968. Ibid. 3: 2 Cryptophyceae, Chloromonadophyceae, Dinophyceae.

Hustedt, F., 1961–66. Die Kieselalgen Deutschlands. Österreichs und der Schweiz unter Berücksichtigung der übrigen Länder Europas sowie der angrenzenden Meeresgebiete. In Kryptogamenflora von Deutschland. Osterreich und der Schweiz (ed. L. Rabenhorst) Bd. 7, Teil 3. Leipzig: Akademische Verlagsgesellschaft Geest & Portig.

Hutchinson, G. E., 1957. A treatise on limnology 1. New York: Wiley.

Hutchinson, G. E., 1967. A treatise on limnology. 2. New York: Wiley.

Jenkin, P. M., 1930. A preliminary limnological survey of Loch Awe (Argyllshire) Part 1: An investigation of some physical and chemical conditions in the loch and experiments on photosynthesis at various depths. Int. Rev. ges. Hydrobiol. Hydrogr. 24: 24–46.

Jørgensen, E. G., 1957. Diatom periodicity and silicon assimilation. Dansk bot. Ark. 18: 1–54.

Knudson, B. M., 1952. The diatom genus *Tabellaria* 1. Taxonomy and morphology. Ann. Bot. N. S. 16: 421–440.

Knudson, B. M., 1953a. The diatom genus *Tabellaria* II. Taxonomy and morphology of the plankton varieties. Ann. Bot. N.S. 17: 131–155.

115

Knudson, M., 1953b. The diatom genus *Tabellaria* III. Problems of infra-specific taxonomy and evolution in *T. flocculosa*. Ann. Bot. N.S. 17: 597–609.

Knudson, B. M., 1954. The ecology of the diatom genus *Tabellaria* in the English lake district. J. Ecol. 42: 345–358.

Knudson, B. M., 1955. The distribution of *Tabellaria* in the English Lake District. Verh. int. Verein. theor. angew. Limnol. 12: 216–218.

Komàrek, J. & Ettl, H., 1958. Algologische Studien. Prague.

Kristiansen, J., 1971. Phytoplankton of two Danish lakes with special reference to seasonal cycles of the nannoplankton. Mitt. int. Verein. theor. angew. Limnol. 19: 253–265.

Lund, J. W. G., 1949. Studies on *Asterionella* I. The origin and nature of the cells producing seasonal maxima. J. Ecol. 37: 389–419.

Lund, J. W. G., 1950a. Studies on *Asterionella formosa* Hass. II. Nutrient depletion and the spring maximum. Part 1. Observations on Windermere, Esthwaite Water and Blelham Tarn. J. Ecol. 38: 1–14.

Lund, J. W. G., 1950b. Studies on *Asterionella formosa* Hass. II. Nutrient depletion and the spring maximum. Part II. Discussion. J. Ecol. 38: 15–35.

Lund, J. W. G., 1954. The seasonal cycle of the plankton diatom *Melosira italica* (Ehr.) Kütz. subsp. *subarctica* O. Müll. J. Ecol. 42: 151–179.

Lund, J. W. G., 1955. Further observations on the seasonal cycle of *Melosira italica* (Ehr.) Kütz. subsp. *subarctica* O. Müll. J. Ecol. 43: 90–102.

Lund, J. W. G., 1959a. A simple counting chamber for nannoplankton. Limnol. Oceanogr. 4: 57–65.

Lund, J. W. G., 1959b. Buoyancy in relation to the ecology of the freshwater phytoplankton. Br. phycol. Bull. 7: 1–17.

Lund, J. W. G., 1961a. The algae of Malham Tarn. Field Studies 1: 85–119.

Lund, J. W. G., 1961b. The periodicity of μ-algae in three English lakes. Verh. int. Verein. theor. angew. Limnol. 14: 147–154.

Lund, J. W. G., 1962a. Concerning a counting chamber for nannoplankton described previously. Limnol. Oceanogr. 2: 261–262.

Lund, J. W. G., 1962b. A rarely recorded but very common British alga, *Rhodomonas minuta* Skuja. Br. phycol. Bull. 2: 133–139.

Lund, J. W. G., 1964. Primary production and periodicity of phytoplankton. Verh. int. Verein. theor. angew. Limnol. 15: 37–56.

Lund, J. W. G., 1965. The ecology of the freshwater phytoplankton. Biol. Rev. 40: 231–293.

Lund, J. W. G., 1971a. The Fritsch collection of illustrations of freshwater algae. Mitt. int. Verein. theor. angew. Limnol. 19: 314–16.

Lund, J. W. G., 1971b. An artificial alteration of the seasonal cycle of the plankton diatom *Melosira italica* subsp. *subarctica* in an English lake. J. Ecol. 59: 521–533.

Lund, J. W. G., 1971c. The seasonal periodicity of three planktonic desmids in Windermere. Mitt. int. Verein. theor. angew. Limnol. 19: 3–25.

Lund, J. W. G., Kipling, C. & LeCren, E. D., 1958. The inverted microscope method of estimating algal numbers and the statistical basis of estimations by counting. Hydrobiologia, 11: 143–170.

Maitland, P. S., 1966. Studies on Loch Lomond 2. Glasgow: Blackie.

Maitland, P. S., 1981. The ecology of Scotland's largest lochs: Lomond, Awe, Ness, Morar and Shiel. Ed. P. S. Maitland. Introduction and catchment analysis. This volume. Chap. 1.

Maitland, P. S., Smith, B. D. & Adair, S. M., 1981a, The ecology of Scotland's largest lochs: Lomond, Awe, Ness, Morar and Shiel. Ed. P. S. Maitland. The fish and fisheries. Chap. 9.

116

This volume.

Maitland, P. S., Smith, B. D. & Dennis, G. M. 1981b. The ecology of Scotland's largest lochs: Lomond, Awe, Ness, Morar and Shiel. Ed. P. S. Maitland. The crustacean zooplankton. Chap. 6. This volume.

Maitland, P. S., Smith, I. R., Bailey-Watts, A. E., George, D. G., Lyle, A. A., Smith, B. D. Duncan, P., Rosie, A. J., Dennis, G. M. & Carr, M. J., 1981c. Twenty-four hour studies of the effects of pumped-storage power stations on water and plankton in Loch Awe (Cruachan) and Loch Ness (Foyers), Scotland. In preparation.

Maitland, P. S., Smith, I. R., Bailey-Watts, A. E., Smith, B. D. & Lyle, A. A., 1981d. The ecology of Scotland's largest lochs: Lomond, Awe, Ness, Morar and Shiel. Ed. P. S. Maitland. Comparisons and Synthesis. Chap. 10. This volume.

Maulood, B. K., & Boney, A. D., 1980. A seasonal and ecological study of the phytoplankton of Loch Lomond. Hydrobiologia, 71: 239–259.

Maulood, B. K., Hinton, G. C. F. & Boney, A. D., 1978. Diurnal variation of phytoplankton in Loch Lomond. Hydrobiologia, 58: 99–117.

Mortimer, C. H., 1942. The exchange of dissolved substances between mud and water in lakes. J. Ecol. 30: 147–201.

Moss, B., 1972. The influence of environmental factors on the distribution of freshwater algae: an experimental study 1. Introduction and the influence of calcium concentration. J. Ecol. 60: 917–32.

Moss, B., 1973a. The influence of environmental factors on the distribution of freshwater algae: an experimental study. II. The role of pH and the carbon dioxide-bicarbonate system. J. Ecol. 61: 157–177.

Moss, B., 1973b. The influence of environmental factors on the distribution of freshwater algae: an experimental study. III. Effects of temperature, vitamin requirements and inorganic nitrogen compounds on growth. J. Ecol. 61: 179–192.

Moss, B., 1973c. The influence of environmental factors on the distribution of freshwater algae: an experimental study. IV. Growth of test spp. in natural lake waters, and conclusion. J. Ecol. 61: 193–211.

Murray, J. & Pullar, L., 1910. Bathymetrical Survey of The Freshwater Lochs of Scotland. Edinburgh: Challenger. Vols. 1–6.

Nygaard, G., 1949. Hydrobiological studies of some Danish ponds and lakes II. The quotient hypothesis and some new or little known phytoplankton organisms. K. danske Vidensk. Selsk. Skr. 7, No. 1.

Ostenfeld, C. H., 1913. De danske farvandes Plankton i aurene 1898–1901. Phytoplankton og Protozoer. K. danske Vidensk. Selsk. Skr. 7. 9, (2): 113–478.

Pavoni, M., 1963. Die Bedeutung des Nannoplanktons im Vergleich zum Netzplankton. Schweiz. Z. Hydrol. 25: 215–341.

Pearsall, W. H., 1932. Phytoplankton of the English Lakes 2. Composition of the phytoplankton in relation to dissolved substances. J. Ecol. 20: 241–262.

Rawson, D. S., 1955. Morphometry as a dominant factor in the productivity of large lakes. Verh. int. Verein. theor. angew. Limnol. 12: 164–175.

Rawson, D. S., 1956. Algal indicators of trophic lake types. Limnol. Oceanogr. 1: 18–25.

Reynolds, C. S., 1973. The phytoplankton of Crose Mere, Shropshire. Br. phycol. J. 8: 153–62.

Ridley, J. E., 1970. The biology and management of eutrophic reservoirs. Wat. Treat. Exam. 19: 374–99.

Ruttner, F., 1963. Fundamentals of limnology. 3rd edition. Transl. by Frey, D. G. and Fry, F. E. J. University of Toronto Press.

117

Ruzicka, J., 1977. Die Desmidiaceen Mitteleuropas Band 1. 1. Stuttgart: E. Schweizerbart'sche.

Slack, H. D., 1957. Studies on Loch Lomond. 1. Glasgow: Blackie.

Small, L. F., 1963. Effect of wind on the distribution of chlorophyll *a* in Clear Lake, Iowa. Limnol. Oceanogr. 8: 426–432.

Smith, B. D., Maitland, P. S., Young, M. R. & Carr, M. J., 1981a. The ecology of Scotland's largest lochs: Lomond, Awe, Ness, Morar and Shiel. Ed. P. S. Maitland. The littoral zoobenthos. Chap. 7. This volume.

Smith, I. R., Lyle, A. A. & Rosie, A. J., 1981b. The ecology of Scotland's largest lochs: Lomond, Awe, Ness, Morar and Shiel. Ed. P. S. Maitland. Comparative physical limnology. Chap. 2. This volume.

Talling, J. F., 1960. Self-shading effects in natural populations of a planctonic diatom. Wett. Leben. 12: 235–242.

Talling, J. F., 1965. Comparative problems of phytoplankton production and photosynthetic productivity in a tropical and a temperate lake. Memorie 1st. ital. Idrobiol. 18 Suppl: 399–424.

Talling, J. F., 1970. Generalized and specialized features of phytoplankton as a form of photosynthetic cover. In: Prediction and Measurement of Photosynthetic Productivity. Proceedings of the IBP/PP Technical Meeting, Trebon 14–21 September 1969. 431–445.

Talling, J. F., 1971. The underwater light climate as a controlling factor in the production ecology of freshwater phytoplankton. Mitt. int. Verein. theor. angew. Limnol 19: 214–43.

Talling, J. F. & Driver, D., 1963. Some problems in the estimation of chlorophyll *a* in phytoplankton. Proc. conf. on Primary Productivity Measurement, Marine and Freshwater, Univ. of Hawaii, Aug 21st-Sept. 6th, 1961. U.S. Atomic Energy Commission, Division of Technical Information TID-7633 (Publ. 1963) pp 142–146.

Tessenow, U., 1964. Untersuchungen über den Kieselsäurehaushalt der Binnengewässer. Arch. Hydrobiol. 32: 1–136.

Tippett, R., 1978. The effect of the Cruachan Pumped-Storage hydro-electric scheme on the limnology of Loch Awe. University of Glasgow: Unpublished report.

Tolstoy, A., 1977. Methods of determining chlorophyll *a* in phytoplankton. Statens Naturvardsverk SNV PM 831 NLU Rapport 91.

Vollenweider, R. A., 1968. Scientific fundamentals of the eutrophication of lake and flowing waters, with particular reference to nitrogen and phosphorus as factors in eutrophication. Tech. Report, Water Management Res. O.E.C.D.

West, W. & West, G. S., 1903. Scottish freshwater plankton – No. 1. Lin. Journ. – Botany, 35: 519–556.

West, W. & West, G. S., 1904/1923. A Monograph of the British Desmidiaceae. London: Ray Society. Vols. 1-5.

West, W. & West, G. S., 1905. Further contribution to the freshwater plankton of the Scottish lochs. Trans. R. Soc. Edinb. 41: 477.

West, W. & West, G. S., 1909. The British freshwater phytoplankton, with special reference to the Desmid-plankton and the distribution of British Desmids. Proc. R. Soc. B, 81: 165–206.

West, W. & West, G. S., 1912. On the periodicity of the phytoplankton of some British lakes. J. Linn. Soc. Bot. XL: 395–432.

Youngman, R. E., 1971. Algal monitoring of water supply reservoirs and rivers. Tech. Mem. Wat. Res. Assoc. TM. 63: 1–26.

Youngman, R. E., 1975. Observations on Farmoor, a eutrophic reservoir in the upper Thames valley during 1965–1973. In The effects of storage on water quality, Medmenham, WRC. 163–202.

5. A review of macrophyte studies

A. E. Bailey-Watts & P. Duncan

Abstract

A brief review is made of both published and unpublished information on the aquatic macrophyte communities of these five waters. Plants generally perform poorly in the large loch environment; near the water surface, exposure to waves has reduced much of the shoreline to a bare rocky substrate, and in deeper water, light is soon attenuated to limiting levels. In localised sheltered areas the physical conditions may be more conducive to plant colonisation; however, even here low levels of nutrients in the oxidised sediments and in the overlying water apparently prevent the development of very extensive stands of vegetation. Floristic information from the few areas studied, indicated that the small tufted species *Isoetes lacustris*, *Lobelia dortmanna*, *Littorella uniflora* and *Subularia aquatica* are the most prevalent species. *Potamogeton* species occur in the comparatively rich (mesotrophic) southern basin of Loch Lomond, but again only in sheltered areas.

Introduction

The aim of this paper is to review the available information on aquatic macrophytes – mainly the submerged vascular hydrophytes, but also dominant species of *Characeae* (Algae) – of five large freshwater lochs in Scotland: Lomond (56°05′ N, 4°35′ W), Awe (56°20′ N, 5°10′ W), Ness (57°16′ N, 4°30′ W), Morar (56°57′ N, 5°45′ W) and Shiel (56°51′ N, 5°30′ W). The work forms part of an integrated limnological study (Maitland 1981) covering physical, chemical and other biological aspects of these waters (Smith *et al.* 1981b, Bailey-Watts & Duncan 1981a, 1981b; Maitland *et al.* 1981b, Smith *et al.* 1981a, Maitland *et al.* 1981a).

This paper differs from the others cited in that no recent field studies were done for it by the authors. Further, it does not aim at a direct comparison of the five lochs, rather a general appraisal of macrophyte work already carried out. In reviewing others' work, however, it completes the overall study by considering a freshwater community which would otherwise have been omitted. Attention to the littoral flora, albeit brief, is also of value to our assessment of the ecological impacts of pumped storage hydro-electric schemes on some of these lochs. This aspect is covered in a report to the North of Scotland Hydro-Electric Board (see also Maitland 1981). In this connection the south basin of Loch Lomond is important and its flora is given specific mention below. In our other papers the observations on Loch Lomond are restricted to the north linear basin.

The status of macrophytes in these lochs

Literature

Studies on the macrophytes of Scottish lochs began early in the present century with the work of West (1905, 1909, both largely paraphrased 1910). Since that time however, only Spence (e.g. 1964) and Britton & Morgan (in Ratcliffe 1977) have investigated the broad distribution of higher aquatic flora in Scotland, although others have collaborated with Spence on various physiological studies, especially those concerned with light relations (see e.g. Spence 1967, 1972, 1976, Spence *et al.* 1971). Idle (e.g. 1968) has studied the littoral and nearby terrestrial flora of parts of Loch Lomond. In contrast to the other lochs which have remained relatively unstudied, Loch Lomond has been the subject of intensive scientific investigation – mainly by the University of Glasgow. Publications by Slack (1957), Weerekoon (in Slack, 1957), Maitland (1966), Tippett (1974), and Boney (1978), comprise a succession of reviews of a number of studies including work on the vascular hydrophytes. Spence has also collected information on aquatic plants in Loch Lomond as well as Lochs Awe and Ness*. There appears to be little information on Lochs Morar and Shiel (*cf.* in Ratcliffe 1977, referred to below).

---

* Professor D. H. N. Spence (University of St Andrews) and Mr E. T. Idle (Nature Conservancy Council, Edinburgh) have each kindly allowed us to quote from their unpublished botanical notes.

Table 5.1 Lochs Lomond, Awe, Ness, Morar and Shiel: physical and chemical features relevant to the macrophyte studies. (Data from Smith *et al.* 1981b; Smith *et al.* 1981a and Bailey-Watts & Duncan 1981a)

Feature	Lomond	Awe	Ness	Morar	Shiel
Length (km)	36	41	39	19	28
Mean breadth (km)	1.95	0.94	1.45	1.42	0.70
Mean depth (m)	37	32	132	87	41
Maximum depth (m)	190	94	230	310	128
Length of shore* (km)	153	129	86	60	85
Shoreline development**	3.44	5.18	3.20	2.18	4.95
Shoreline mean gradient	9.91[a] 28.78[b]	14.09	3.71	9.92	15.97
Light conditions:					
Secchi disc readings (m)	4.20-6.75	2.20-4.20	3.60-4.60	5.75-10.20	5.10-8.00
Hazen colour[a]	5-15	15-40	5-25	5-10	5-10
Colour and 1% depth (m) of wavebands showing					
(a) most penetration	green 6.25	red 4.30	red 4.65	orange 9.50	orange 8.35
(b) least penetration	blue 2.20	blue 1.20	blue 1.30	blue 3.35	blue 2.20
pH (annual mean)[a]	6.78	6.90	6.70	6.63	6.14
Alkalinity (mEq CaCO$_3$ l^{-1})[a]	0.13	0.18	0.09	0.07	0.04

* including islands.   a North basin
** excluding islands.   b South basin

General botanical features in relation to the large loch environment

Physical and chemical features of the lochs, of relevance to a consideration of their littoral floras, are presented in Table 5.1. Fig. 5.1-5.3 are outline maps of Lochs Lomond, Awe and Ness, with the locations of sampling areas mentioned later in this review.

Although the amount of data on the littoral macrophytes of these lochs is limited, it is sufficient to substantiate the view that growth of both emergent and submerged species is generally sparse. Moreover, the early observations of West (1905, 1910) on e.g. Loch Ness, with which Spence (1964) compared his own findings, show that this situation has remained relatively unchanged there and in certain other locations over 56 years. The essentially rocky substrates (Plate 5.1) of the littoral areas of these large exposed waters permit little colonisation by higher plants, although cryptogamic elements (micro-algae and bryophytes) may form an epilithic cover (Slack 1957, Spence 1967). Even in sheltered bays (Plate 5.2), where a form of organic soil can develop, production of higher aquatics is limited (see also Ratcliffe 1977); muds of these lochs are generally brown with an oxidised surface layer through which

121

Plate 5.1 Exposed beaches with stony and rocky substrates typical of much of the shoreline of the large Scottish lochs (a) Loch Lomond south of Tarbet, (b) Loch Ness at Dores,

122

(c) Loch Morar at Bracora and (d) Loch Shiel near Glenfinnan.
(Photos: B. D. Smith & P. S. Maitland.)

123

Plate 5.2   Sheltered bays of Loch Awe showing localised stands of emergent vegetation. (a) and (b) silty areas near Kilchurn,

124

(c) sandy area north of Kilchurn and (d) silty area at south end.
(Photos: B. D. Smith.)

125

little exchange of plant nutrients takes place. In addition, the level of nutrients in the overlying water is extremely low in these oligotrophic lakes (Spence 1967, see also Bailey-Watts & Duncan 1981a).

Whilst substrates more conducive to rooted plant colonisation can occur in deeper water, i.e. outside the zone dominated by the eroding power of waves, at such depths the light necessary for photosynthesis is often extremely attenuated and this limits plant growth. This is especially the case in the peatier waters of Lochs Ness and Awe which contrast with the other lochs in their higher Hazen colour values, lower Secchi disc readings and light extinction coefficients (Table 5.1 and see Bailey-Watts & Duncan 1981a). Slack (1957) takes the uppermost 4 m zone as potentially suitable for higher plant growth in Loch Lomond. He calculates that this 'phytal' zone covers 8% of the northern 'Highland' stretch of the loch, and 11% of the southern 'Lowland' basin. Spence (1964) follows West's (1910) suggestion that the photic zone of e.g. Loch ness is approximately 30′ (9.2 m). Disregarding sheltered bays because they represent less than 0.1% of the area of Loch Ness, Spence calculated (from Murray & Pullar 1910) that only 5% of the loch area lies between 0 and 9.15 m. Assuming that of this, only 5% is actually covered by plants, he considered the likely colonised area to be less than 0.25%. Similar calculations indicated equivelent values of around 1% for Loch Lomond, and 0.7% for Loch Morar.

The present observations, and the records of Smith *et al.* (1981a) (and B. D. Smith personal communication), suggest that plant growth is negligible along vast lengths of shore-line of the five lochs visited for our other studies. In most of them, extremely deep water is reached within only a few metres of the shore – see also Murray & Pullar (1910), West (1905), Spence (1967). Table 5.1 gives information on the shorelines of these deep and narrow fjord- like lochs (cf. Loch Lomond south basin).

Thus significant stands of either reed swamp or submerged species occur only in sheltered areas such as Urquhart and Inchnacardoch Bays in Loch Ness (West 1905, 1910) and various sheltered localities in Lochs Morar and Shiel (Britton & Morgan in Ratcliffe 1977) and Lochs Awe and Lomond (Slack 1957, Spence 1964 and unpublished; Idle, unpublished). Furthermore, Idle's notes for the island and south-eastern mainland shores (Fig. 5.1) of Loch Lomond (that area bounded by the Loch Lomond National Nature Reserve – see also Idle 1978) indicate few substantial stands of vegetation. Exposure to wave action has resulted in a poorly colonised rocky terrace around many of the islands. Sand bars, moulded by water circulation asso-ciated with the River Endrick at its point of entry to the loch, have also produced a relatively unstable platform for plant growth.

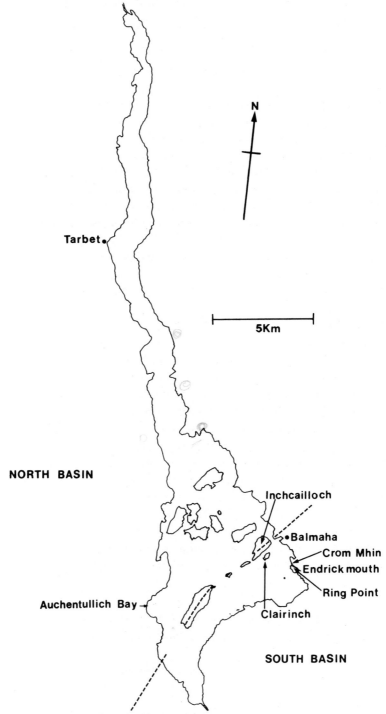

N

Tarbet●

5Km

NORTH BASIN

Inchcailloch

●Balmaha
Crom Mhin
Endrick mouth

Auchentullich Bay

Ring Point

Clairinch

SOUTH BASIN

Fig. 5.1 Outline map of Loch Lomond showing sampling sites and other areas mentioned in the text. Dotted line marks the line of the Highland Boundary Fault.

127

Plant associatons at selected shore and near-shore sites

Species recorded at various localities along the shores of these lochs are mainly of the small tufted plant type e.g. *Isoetes lacustris, Lobelia dortmanna, Subularia aquatica* and *Littorella uniflora*. Apart from the fairly ubiquitous *Littorella*, these are forms typical of Spence's (1967) poor (often brown) water class of loch with alkalinity levels of less than 0.3 mEq l⁻¹. Mean alkalinity levels for the 1978 period covered by Bailey-Watts & Duncan's (1981a) study, were from 0.18 mEq l⁻¹ in Loch Awe down to 0.04 mEq l⁻¹ in Loch Shiel. Their work on Loch Lomond did not encompass the richer south basin, where alkalinities can often exceed 0.3 mEq $CaCO_3$ l⁻¹. Idle quotes unpublished data from the Clyde River Purification Board of up to 0.8 mEq l⁻¹ near Balmaha and the River Endrick.

Loch Lomond

Our consideration of the botanical lists from these lochs begins with the comparatively rich south basin of Loch Lomond. Table 5.2 contains the species recorded by Idle (unpublished data from summer 1967) in his transect zones 3 and the permanently submerged zones. *Littorella uniflora*, and *Isoetes lacustris* were recorded at all of the 8 sites, *Myriophyllum alterniflorum* was recorded at all sites except 'Clairinch 5 and 9'. The charophyte *Nitella opaca* was not recorded At 'Clairinch 9' and 'Inchcailloch 10'.

This study confirms the impressions of Spence (1964) that *Isoetes* is often one of few dominants, or the sole species found at depth in large Scottish lochs. However, the charophyte *Nitella opaca* may sometimes extend to deeper areas than the vascular hydrophytes (West 1905, Spence 1964).

Spence summarises the horizontal distribution of species on the east shore of the Loch Lomond south basin, with the following scheme:

Inshore:	*Carex vesicaria* or *Eleocharis* or *Glyceria fluitans*
	Polygonum amphibium or *G. fluitans*
	Littorella
Offshore:	Sparse *Elatine* on bottom sand

Substrate – flora associations at Auchentullich Bay (south basin of Loch Lomond) studied by Weerekoon (see Slack 1957) were as follows:

Table 5.2 Dominant plant species recorded in 1967 by Idle (unpublished) in the submerged zones and zones 3 at various shore locations around Clairinch and Inchcailloch and Ring Point and Crom Mhin on the south-east mainland shore of Loch Lomond

Species	Clairinch				Inchcailloch		Ring Point	Crom Mhin
	5	6	7	9	5	10		
Submerged zones								
Eleocharis acicularis	−	−	−	−	−	−	+	−
E. palustris	−	−	−	−	−	−	+	−
Isoetes lacustris	+	+	+	+	+	+	+	+
Littorella uniflora	+	+	+	+	+	+	+	+
Myriophyllum alterniflorum	−	+	+	−	+	+	+	+
Nitella opaca	+	+	+	−	+	−	+	+
Potamogeton crispus	−	−	−	−	−	−	+	+
P. × nitens	−	−	−	−	−	−	−	+
P. obtusifolius	−	−	−	−	−	−	−	+
P. perfoliatus	−	−	−	−	−	−	+	+
Ranunculus flammula	−	−	−	−	+	−	−	−
R. heterophyllea	−	−	−	−	−	−	−	+
Zone 3								
Alnus glutinosa	+	−	−	−	−	−	−	−
Caltha palustris	−	−	+	−	−	−	−	−
Carex nigra	−	−	−	+	−	−	−	−
C. acutiflorus	−	−	−	+	−	−	−	−
Carum verticillatum	−	−	−	+	+	−	−	−
Deschampsia caespitosa	−	−	−	−	−	+	−	−
Eleocharis acicularis	−	−	−	−	−	−	+	+
E. palustris	−	−	+	−	−	−	+	−
Equisetum fluviatile	−	−	+	−	−	−	+	+
Fontinalis antipyretica	+	+	−	+	−	+	−	−
Hydrocotyle vulgaris	+	−	−	−	−	−	−	−
Juncus acutiflorus	+	+	−	+	−	−	−	−
J. bufonius	−	−	−	−	−	−	+	−
J. fluitans	−	−	+	−	−	−	−	−
J. articulatus	+	−	−	−	−	−	−	−
Littorella uniflora	+	−	+	+	+	+	−	+
Lobelia dortmanna	−	−	+	−	−	−	−	−
Molinia caerulea	+	+	−	+	+	+	−	−
Myosotis scorpioides	−	−	−	−	−	−	−	+
Myriophyllum alterniflorum	−	−	+	−	−	−	−	−
Phalaris arundinacea	+	+	−	−	−	+	−	−
Polygonum hydropiper	−	−	−	−	−	−	−	+
Ranunculus flammula	+	+	−	+	−	−	−	−
Salix ? atrocinarea	−	−	−	−	−.	−	+	−

Marginal cobbles, sand and scattered stone	'Aufwuchs' of epilithic diatoms and *Fontinalis*
Sand at 1 m	*Littorella uniflora* and rosettes of *Lobelia dortmanna*
Finer silts	*Isoetes lacustris* and, in more sheltered areas, *Myriophyllum alterniflorum*
Muds	*Nitella opaca*

Investigations at a shore approximately 5 km north of Tarbet in the northern basin revealed only sparse *Littorella* and occasional *Isoetes* in gaps between stones and rocks which extended to water depths of 1.1-1.2 m (Spence, unpublished). With many stones still present at 2.0 m, *Isoetes* was dominant, with *Myriophyllum*, *Lobelia* and *Littorella* less common, and *Juncus fluitans* rare. Although below 2m the shore at this point flattens, the muds there support only *Isoetes*.

Loch Awe

At a bay on the east side of the north end of Loch Awe, opposite Kilchurn Castle (Fig. 5.2 and see Plate 5.2), Spence recorded an assemblage of submerged species essentially similar to that described above. At 70-80 cm water depth *Littorella, Lobelia, Isoetes,* and *Subularia aquatica* were prominent, and the last two, along with *Elodea canadensis*, extended to at least 1.9 m depth. Other species are included in Table 5.3.

Loch Ness

At Inchnacardoch and Urquhart Bays in Loch Ness (Fig. 5.3), a different assemblage of species occurred (Table 5.4). However, the deepest water in which the records were made was only 55 cm and the flora present in the shallows (5 and 10 cm) included marsh forms.

Loch Morar

In sheltered areas of Loch Morar, the previously mentioned commonplace association of *Littorella, Lobelia* and *Isoetes* is found (Britton & Morgan in Ratcliffe 1977). Here too *Isoetes* alone may form locally pure and dense swards to about 5 m depth.

130

Table 5.3 Submerged vascular species recorded at four points along the littoral zone of a bay in Loch Awe at the east side of the north end, opposite Kilchurn Castle (data from Spence, unpublished)

Water depth (cm)	Species
40	*Carex vesicaria*
	Juncus fluitans
70	*Equisetum fluviatile*
	Isoetes lacustris
	Eleocharis palustris
	Elodea canadensis
	J. fluitans
	Schoenoplectus lacustris
	Nuphar pumila
	Littorella uniflora
	Lobelia dortmanna
	Ranunculus flammula
	Subularia aquatica
80	*J. fluitans*
	S. lacustris
	Sparganium minimum
	Callitriche hermaphroditica
	L. uniflora
	L. dortmanna
190	*I. lacustris*
	E. canadensis
	S. aquatica

Table 5.4 Species recorded by Spence (unpublished) from two bays on Loch Ness

Water depth (cm)	Species
Inchnacardoch Bay	
10	*Scutellaria galericum*
	Veronica scutellata
50	*Equisetum fluviatile*
	Carex rostrata
	Littorella uniflora
	Menyanthes trifoliata
Urquhart Bay	
5	*Mentha aquatica*
	Menyanthes trifoliata
	Polygonum amphibium
	Potentilla palustris
55	*Myriophyllum alterniflorum*
	(Fontinalis antipyretica)

131

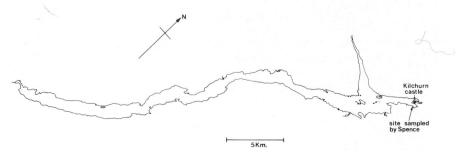

Fig. 5.2 Outline map of Loch Awe showing the site sampled by Spence and referred to in the text.

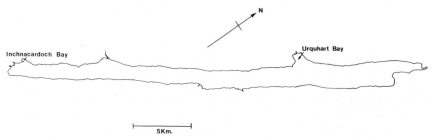

Fig. 5.3 Outline map of Loch Ness showing the position of the two bays referred to in the text.

Loch Shiel

Britton & Morgan (loc. cit.) make the same general remarks regarding the submerged macrophyte community in Loch Shiel as in Loch Morar.

General

The sparse plant assemblages referred to above, thin out rapidly as the upshore zone of inundation is traversed. Essentially terrestrial species are then found e.g. island woodlands on Loch Morar (McVean 1958). Their abundance and species composition depends on the type of shore, its slope and aspect etc. In sheltered areas again, a swamp or marsh-like zone may develop, and a few limited areas of this type may be found around all the lochs. Along the greater length of most of them however, the zone of semi-emergent or stranded forms gives way quickly to terrestrial species. Periodically inundated grassy areas may occur, but otherwise the 'wet' and essentially 'dry' zones are separated by a band of virtually bare stone and rocks. On many of the Loch Lomond island shores studied by Idle, this band is very

132

narrow; depending on loch level, wind and wave action, only a few metres separate severely exposed tree roots and the woodland edge, from the water edge. At the other extreme lie the more gently sloping areas around the River Endrick mouth. Here, especially during prolonged dry weather, extensive areas of sand and mud are exposed. If such conditions coincide with the summer growing period of many plants, the areas may become rapidly colonised. During the dry summer of 1968, Idle, Mitchell & Stirling (1970) recorded *Elatine hydropiper* L. (new to Scotland) on these mud flats. The related *E. hexandra* (Lapiere) DC had been recorded at Balmaha in 1953 following a similar period of dry weather and low water levels.

Acknowledgements

Especial thanks are due to Prof. D. H. N. Spence and Mr E. T. Idle for putting their unpublished data at our disposal. Professor Spence also commented on an earlier version of this paper. We are grateful to them and to Mr John Mitchell (Nature Conservancy Council) for showing us round the areas of Loch Lomond referred to in our review.

References

Bailey-Watts, A. E. & Duncan, P. 1981a. The ecology of Scotland's largest lochs: Lomond, Awe, Ness, Morar and Shiel. Ed. P. S. Maitland. Chemical characterisation – a one-year comparative study. Chap. 3. This volume.

Bailey-Watts, A. E. & Duncan, P. 1981b. The ecology of Scotland's largest lochs: Lomond, Awe, Ness, Morar and Shiel. Ed. P. S. Maitland. The phytoplankton. Chap. 4. This volume.

Boney, A. D. 1978. Microscopic plant life in Loch Lomond. Glasg. Nat. 19: 391–402.

Idle, E. T. 1968. *Rumex aquaticus* L., at Loch Lomondside. Trans. Proc. bot. Soc. Edinb. 40: 445–449.

Idle, E. T. 1978. The flora of the Loch Lomond National Nature Reserve. Glasg. Nat. 19: 403–421.

Idle, E. T. 1967. A preliminary analysis of the relationship between physical and biotic factors and the vegetation of some shores of Loch Lomond. Unpublished report.

Idle, E. T., Mitchell, J. & Stirling, A. McG. 1970 *Elatine hydropiper* L., new to Scotland. Watsonia, 8: 45–46.

McVean, D. N. 1958. Island vegetation of some West Highland freshwater lochs. Trans. Proc. bot. Soc. Edinb. 37: 200–208.

Maitland, P. S. 1966. Studies on Loch Lomond 2. Glasgow: Blackie.

Maitland, P. S. 1981. The ecology of Scotland's largest lochs: Lomond, Awe, Ness, Morar and Shiel. Ed. P. S. Maitland. Introduction and catchment analysis. Chap. 1. This volume.

Maitland, P. S., Smith, B. D. & Adair, S. M. 1981a. The ecology of Scotland's largest lochs: Lomond, Awe, Ness, Morar and Shiel. Ed. P. S. Maitland. The fish and fisheries. Chap. 9. This volume.

Maitland, P. S., Smith, B. D. & Dennis, G. M. 1981b. The ecology of Scotland's largest lochs: Lomond, Awe, Ness, Morar and Shiel. Ed. P. S. Maitland. The crustacean zooplankton. Chap. 6. This volume.

Murray, J. & Pullar, L. 1910. Bathymetrical survey of the freshwater lochs of Scotland. Edinburgh: Challenger. Vols. 1–6.

Ratcliffe, D. A. 1977. A nature conservation review. 1. Cambridge University Press.

Slack, H. D. 1957. Studies on Loch Lomond I. Glasgow: Blackie.

Smith, B. D., Maitland, P. S., Young, M. R. & Carr, M. J. 1981a. The ecology of Scotland's largest lochs: Lomond, Awe, Ness, Morar and Shiel. Ed. P. S. Maitland. The littoral zoobenthos. Chap. 7. This volume.

Smith, I. R., Lyle, A. A. & Rosie, A. J. 1981b. The ecology of Scotland's largest lochs: Lomond, Awe, Ness, Morar and Shiel. Ed. P. S. Maitland. Comparative physical limnology. Chap. 2. This volume.

Spence, D. H. N. 1964. The macrophytic vegetation of freshwater lochs, swamps and associated fens. In: The vegetation of Scotland (ed. J. H. Burnett). Edinburgh: Oliver & Boyd. 306–425.

Spence, D. H. N. 1967. Factors controlling the distribution of freshwater macrophytes with particular reference to the lochs of Scotland. J. Ecol. 55: 147–170.

Spence, D. H. N. 1972. Light on freshwater macrophytes. Trans. Proc. Bot. Soc. Edinb., 41: 491–505.

Spence, D. H. N. 1976. Light and plant response in freshwater. In: Light as an ecological factor II. Symp. Br. Ecol. Soc. (ed. G. C. Evans, R. Bainbridge, & O. Rackham,) Oxford: Blackwell. 93–133.

Spence, D. H. N. 1960. Botanical notes on Lochs Lomond, Ness and Awe. Unpublished.

Spence, D. H. N., Campbell, R. M. & Chrystal, J. 1971. Spectral intensity in some Scottish freshwater lochs: Freshwat. Biol. 1: 321–337.

Tippett, R. 1974. Life in freshwater. In: A history of Loch Lomond. University of Glasgow Press, 63–73.

West, G. 1905. A comparative study of the dominant phanerogamic and higher cryptogamic flora of aquatic habit, in three lake areas of Scotland. Proc. R. Soc. Edinb. 25: 967–1023.

West, G. 1909. A further contribution to a comparative study of the dominant phanerogamic and higher cryptogamic flora of aquatic habit in Scottish lakes. Proc. R. Soc. Edinb. 30: 65–182.

West, G. 1910. An epitome of a comparative study of the dominant phanerogamic and higher cryptogamic flora of aquatic habit, in seven lake areas of Scotland. Bathymetrical survey of the freshwater lochs of Scotland I. (ed. J. Murray & L. Pullar) Edinburgh: Challenger. 156–260.

6. The crustacean zooplankton

P. S. Maitland, B. D. Smith & G. M. Dennis

Abstract

The crustacean zooplankton of these large lochs is described from a series of
vertical net hauls taken there during 1977–78. A total of 11 planktonic species
of Crustacea was found – six of them common to all the lochs. There were
sufficient differences in the occurrence of the other five species for it to be
possible to identify the community of each loch from a single haul during
summer. The basic population dynamics of all species are described and
compared among these lochs. The lochs themselves are compared on the
basis of a number of features of the crustacean zooplankton – especially the
total number of species and individuals, and the mean and maximum numbers
of each species. It was found that Loch Lomond exhibited an overall dominance in terms of the criteria considered, and that Loch Morar was lowest in
almost all respects. The significance of these findings is discussed in relation
to other factors.

Introduction

This paper is one of a series of studies on the comparative ecology of five of
Scotland's largest freshwater lochs: Lomond, Awe, Ness, Morar and Shiel.
The series reports on a multidisciplinary project carried out from 1977–79 to
study basic aspects of the ecology of these large bodies of water, about which
(other than Loch Lomond) relatively little is at present known. Other papers
in the series are as follows: introduction and catchment studies (Maitland
1981), physics (Smith *et al.* 1981b), chemistry (Bailey-Watts & Duncan
1981a), phytoplankton (Bailey-Watts & Duncan 1981b), macrophytes
(Bailey-Watts & Duncan 1981c), zoobenthos (Smith *et al.* 1981a), and fish

(Maitland *et al.* 1981b). A short-term study of the effects of pumped-storage electrical power stations on two of these lochs (Awe and Ness) forms part of the same project (Maitland *et al.* 1981c), but is not reported in the present series.

Ecological studies at Loch Lomond have been continuing for a number of years (Lamond 1931, Slack 1957, 1965, Chapman 1969, Maitland 1969 etc.) and this is now not only the largest area of fresh water in Great Britain but one of the best known. Recently, interest in this important water body has intensified following the proposal by the North of Scotland Hydro-Electric Board (1976) to build a large pumped-storage powe scheme there at Craig-royston beside the north basin of the loch. The basis of the multidisciplinary project, of which this study forms a part, was a comparison of various aspects of the ecology of Loch Lomond with that of two large lochs with existing pumped-storage schemes (Awe and Ness) and two large lochs without such schemes (Morar and Shiel). Zooplankton at Loch Lomond have been studied previously by Scott (1899), Murray & Pullar (1910), and Chapman (1965, 1969, 1972). However, other than the original studies by Murray (1905), and Murray & Pullar (1910), there has been little work on the plankton of the other lochs. Exceptions are the general surveys of Scott (1901) and Murray & Pullar (1910), and work by Murray (1904) on Loch Morar and by Tippett (1978) on Loch Awe.

Bathymetric surveys of Lochs Lomond, Awe, Ness, Morar and Shiel were carried out at the beginning of this century by Murray & Pullar (1910). A comparison of the major features of their morphometry and chemistry is given in Table 6.1: further physico-chemical and biological data are given in

Table 6.1 Basic physical and chemical features of Lochs Lomond, Awe, Ness, Morar and Shiel relevant to the zooplankton study.

Feature	Lomond	Awe	Ness	Morar	Shiel
National Grid Reference	263598	270017	285429	177690	178072
Surface area (km^2)	71.1	38.5	56.4	25.7	19.6
Catchment area (km^2)	781	840	1775	168	248
Mean depth (m)	37.0	32.0	132.0	86.6	40.5
Maximum depth (m)	189.9	93.6	229.8	310.0	128.0
Volume (m^3 10^8)	26.279	12.304	74.519	23.073	7.925
Catchment: % base richness	45	41	13	0	0
Catchment: % arable	17.0	2.2	1.9	0.7	1.4
Catchment: house nos.	421	123	547	54	118
pH (annual mean)	6.78	6.90	6.70	6.63	6.14
Alkalinity (mg/l CaCO$_3$)	6.37	8.97	4.48	3.27	2.10
Conductivity (K$_{20}$ μS cm^{-1})	34	41	30	35	29

136

the papers cited above. Though all the lochs are large with deep narrow basins and mainly mountainous catchments, there are important differences, particularly in the relative size and quality of the catchments and the hydrology and water chemistry, which result from these.

Methods

This study of the crustacean zooplankton is based on a series of vertical net hauls taken at a single station in each loch at regular intervals from November 1977 to October 1978. Studies of the horizontal distribution of the zooplankton were carried out independently over the same period, and are described elsewhere (George & Jones 1981). Full details of the stations (at which temperature profiles, water chemistry and phytoplankton were also studied coincidently) are given elsewhere by Smith, *et al.* (1981b). Comparative transects across each loch through the stations concerned are shown in Fig. 6.1.

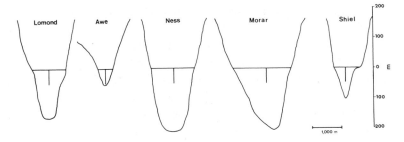

Fig. 6.1 Horizontal sections across the large lochs at the zooplankton sampling stations. The vertical scale is indicated by the 50 m line below the water surface, representing the distance through which the vertical hauls were taken.

During the year of study regular visits were made to all five lochs, usually sampling one loch on each day of a week, starting at Loch Ness then Morar, Shiel, Awe and finally Lomond. Sampling was carried out from a small boat and on each occasion five vertical hauls, from 50 m to the surface, were taken one after the other with a zooplankton net. Occasionally, sampling was impossible on one or more of the lochs because of bad weather, which was also sometimes responsible for fewer than five sets of hauls being taken or occasional hauls shorter or longer than 50 m.

The net used was designed specifically for this study, but followed the criteria recommended by UNESCO (1968). It had a total length of 100 cm,

137

with a circular opening of 28 cm in diameter (Plate 6.1). At the end of the net cone was fixed a detachable phosphor-bronze filter in which animals trapped in the net were concentrated. The mesh aperture of the net was 0.25 mm and of the filter was 0.1 mm (McGowan & Fraundorf 1966). Within the tube which housed the filter was loaded a 2 kg brass weight.

In use, the net was lowered to a depth of 50 m and then hauled steadily to the water surface. Plankton were then washed into the filter by plunging the sides into the water and allowing the net to drain. The filter containing the plankton was then removed and placed in a labelled jar containing 4% formaldehyde. It was replaced by a new filter and another sample taken in the same way. In the laboratory, zooplankton were washed out of the filters into labelled tubes, and subsequently identified and counted.

Results

The communities of crustacean zooplankton

The annual mean and the maximum numbers of the dominant species (Fig. 6.2) in each loch are shown in Tables 6.2 and 6.3, while the seasonal variation in total numbers is given in Fig. 6.3a. Though all the lochs had several species in common (*Bosmina coregoni, Polyphemus pediculus, Bythotrephes longimanus, Leptodora kindti, Diaptomus gracilis* and *Cyclops abyssorum*) a number of other species occurred in only one (*Diaptomus laciniatus, Cyclops leuckarti*), three (*Daphnia hyalina*) or four (*Diaphanosoma brachyurum, Holopedium gibberum*) lochs. Thus in considering only the dominant species there are sufficient characteristics to distinguish the plankton of each loch. In addition, there were some notable differences in the total numbers of various species in the different lochs and in their seasonal variation.

Odd specimens of littoral species of Crustacea were found in the plankton. These were mainly *Alona affinis* Leydig and unidentified Harpacticoida. Nauplius larvae were common in all lochs, especially during summer. Various species of Rotifera occurred in all lochs, sometimes in large numbers. Hydracarina occurred infrequently. Larvae of *Chaoborus flavicans* (Meigen) abundant in an associated lochan (Goldspink & Scott 1971) were found regularly in Loch Lomond. In addition, various occasional and sometimes surprising invertebrates occurred in the net hauls (e.g. *Simulium* in Loch Awe) and in one sample (from Loch Shiel) a small fish (*Gasterosteus aculeatus* Linnaeus) was found. None of the casual occurrences of invertebrates are considered further in this account.

Plate 6.1   Sampling zooplankton at Loch Lomond with the large net used for taking vertical hauls. (Photo: A. J. Rosie.)

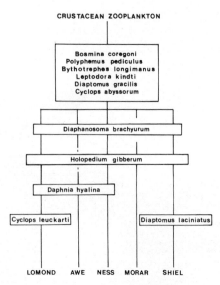

CRUSTACEAN ZOOPLANKTON

Bosmina coregoni
Polyphemus pediculus
Bythotrephes longimanus
Leptodora kindti
Diaptomus gracilis
Cyclops abyssorum

Diaphanosoma brachyurum

Holopedium gibberum

Daphnia hyalina

Cyclops leuckarti

Diaptomus laciniatus

LOMOND   AWE   NESS   MORAR   SHIEL

Fig. 6.2   The characterisation of the five large lochs by their crustacean zooplankton species.

A diagrammatic analysis of the species characterisation of the crustacean zooplankton of the five lochs is shown in Fig. 6.2. Loch Lomond contains ten of the common species but is apparently unique among these lochs in containing *Cyclops leuckarti,* which is especially common during the summer months. Loch Awe contains eight of the common species but does not appear to contain *Holopedium gibberum* (though see below) nor any unique species. Loch Ness contains nine of the common species but nothing else. Loch Morar has the least diverse zooplankton with only seven of the common species and nothing unique. Loch Shiel includes nine of the common species and was unique in the presence during the ummer months of *Diaptomus laciniatus.*

The mean (weighted) numbers of the common species in all the lochs for the year of study (November 1977 to October 1978) are given in Table 6.2, and the maximum numbers per individual haul in Table 6.3. Of the six Crustacea which occurred in all lochs the highest densities of *Polyphemus pediculus* and *Bythothrephes longimanus* were found in Loch Lomond, of *Cyclops abyssorum* in Loch Awe, of *Leptodora kindti* in Loch Ness and of *Bosmina coregoni* and *Diaptomus gracilis* in Loch Shiel (Table 6.2). Only in Loch Morar were no maxima found. A similar situation occurred with the maximum individual hauls of these six species in the five lochs (Table 6.3). Table 6.4 indicates the precedence of the lochs in terms of mean and maximum numbers of these six species.

140

Table 6.2 Mean numbers per m³ of the dominant crustacean zooplankton in Lochs Lomond, Awe, Ness, Morar and Shiel from November 1977 to October 1978

Species	Lomond	Awe	Ness	Morar	Shiel
Diaphanosoma brachyurum Lieven	1	152	1	—	65
Holopedium gibberum Zaddach	30	—	3	3	4
Daphnia hyalina Leydig	170	233	171	—	—
Bosmina coregoni Baird	78	134	78	67	203
Polyphemus pediculus (Linnaeus)	1	1	1	1	1
Bythotrephes longimanus Leydig	3	1	1	1	1
Leptodora kindti (Focke)	2	2	3	1	1
Diaptomus laciniatus Lilljeborg	—	—	—	—	28
Diaptomus gracilis Sars	613	522	341	287	1064
Cyclops abyssorum Sars	73	316	192	17	264
Cyclops leuckarti (Claus)	24	—	—	—	—
Others	6	4	2	1	2
Totals	1101	1365	793	378	1633

Table 6.3 Maximum numbers per m³ of the dominant crustacean zooplankton species occurring in individual vertical hauls in Lochs Lomond, Awe, Ness, Morar and Shiel from November 1977 to October 1978

Species	Lomond	Awe	Ness	Morar	Shiel
Diaphanosoma brachyurum	10	785	1	—	777
Holopedium gibberum	267	—	28	18	32
Daphnia hyalina	3139	558	745	—	—
Bosmina coregoni	269	1076	309	199	712
Polyphemus pediculus	20	16	8	8	3
Bythotrephes longimanus	18	4	8	5	3
Leptodora kindti	18	9	16	8	8
Diaptomus laciniatus	—	—	—	—	184
Diaptomus gracilis	2127	1238	1166	1034	3697
Cyclops abyssorum	327	1756	1730	44	1000
Cyclops leuckarti	80	—	—	—	—

Table 6.4 Precedence of mean (and maximum) numbers of the six species of crustacean zooplankton common to all the large lochs

Species	Lomond	Awe	Ness	Morar	Shiel
Bosmina coregoni	4 (4)	2 (1)	3 (3)	5 (5)	1 (2)
Polyphemus pediculus	1 (1)	2 (2)	4 (3)	3 (4)	5 (5)
Bythotrephes longimanus	1 (1)	2 (4)	3 (2)	4 (3)	5 (5)
Leptodora kindti	2 (1)	3 (3)	1 (2)	5 (5)	4 (4)
Diaptomus gracilis	2 (2)	3 (3)	4 (4)	5 (5)	1 (1)
Cyclops abyssorum	4 (4)	1 (1)	3 (2)	5 (5)	2 (3)
Totals	14 (13)	13 (14)	18 (16)	27 (27)	18 (20)

Diaptomus gracilis (Plate 6.2) was outstanding in dominance for it had the highest mean numbers of any species in all five lochs. *Bosmina coregoni* and *Cyclops abyssorum* also had high numbers in all lochs as did *Daphnia hyalina* in the three lochs in which it occurred and *Diaphanosoma brachyurum* in Lochs Awe and Shiel. The three predatory species (*Polyphemus pediculus*, *Bythotrephes longimanus* and *Leptodora kindti*), though they occurred in all lochs, were never found in high numbers. The mean numbers of *Diaptomus laciniatus* (unique to Loch Shiel) and *Cyclops leuckarti* (unique to Loch Lomond) were relatively low, as were those of *Holopedium gibberum*.

The six species common to the five lochs showed differences in the timing of their first (and usually major) peaks of abundance. The precedence of these is summarised in Table 6.5 It can be seen that Loch Lomond is clearly the 'earliest' of the lochs and Lochs Morar and Shiel the 'latest'. Lochs Awe and Ness occupy an intermediate position between these.

Table 6.5 Precedence of the first peaks of abundance in the numbers of the six species of crustacean zooplankton common to all the large lochs

Species	Lomond	Awe	Ness	Morar	Shiel
Bosmina coregoni	1	1	2	2	2
Polyphemus pediculus	1	1	2	3	2
Bythotrephes longimanus	1	2	1	1	1
Leptodora kindti	1	2	2	3	3
Diaptomus gracilis	1	1	2	2	2
Cyclops abyssorum	1	1	2	2	3
Totals	6	8	11	13	13

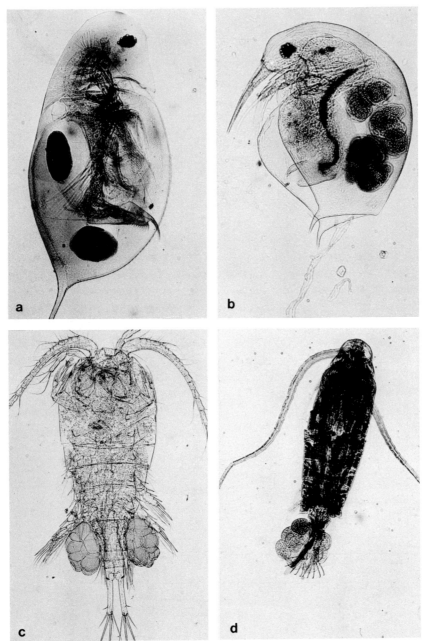

Plate 6.2 Common crustacean zooplankton from Loch Lomond, but species found in most or all of the large lochs. (a) *Daphnia hyalina* (b) *Bosmina coregonia* (c) *Cyclops abyssorum* (d) *Diaptomus gracilis* (Photos: M. A. Chapman.)

143

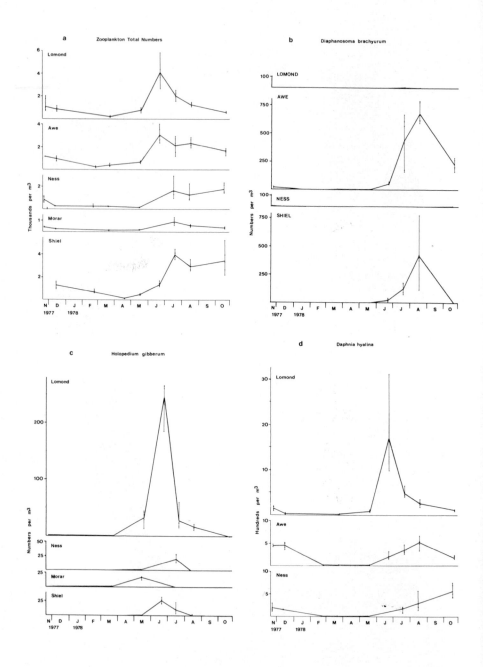

Fig. 6.3 The seasonal abundance in the large lochs from November 1977 to October 1978 of: (a) Total crustacean zooplankton, (b) *Diaphanosoma brachyurum*, (c) *Holopedium gibberum*, and (d) *Daphnia hyalina*. The values associated with each mean are the actual densities in each haul.

The dominant species

Comparative seasonal graphs for the eleven dominant species are given in Figs. 6.3-5. The results for the individual net hauls and the means of each date are included. It is a characteristic feature of almost all the samples that between-sample variation is greatest at the times of greatest densities and vice versa.

Diaphanosoma brachyurum occurred in Lochs Lomond, Awe, Ness and Shiel (Fig. 6.3b) but not in Loch Morar. It was uncommon in Lochs Lomond and Ness but was abundant in Lochs Awe and Shiel with maximum individual samples densities of 785 and 777 individuals per m^3 and mean annual densities of 152 and 65 per m^3 respectively. A similar seasonal cycle occurred in both these lochs, with very low numbers during the winter and spring which rose rapidly from May onwards to reach a maximum in August. Numbers declined rapidly after this.

Holopedium gibberum was found in Lochs Lomond, Ness, Morar and Shiel (Fig. 6.3c), but not in Loch Awe. It was particularly abundant in Loch Lomond with a maximum individual sample density of 267 individuals per m^3 and a mean annual density of 30 per m^3. The seasonal cycle was similar in all the lochs with numbers very low or absent during the winter and spring, but starting to rise in March and April to reach maxima in May (Morar), June (Lomond and Shiel) or July (Ness). Numbers declined rapidly after the maxima had been reached.

Daphnia hyalina occurred in Lochs Lomond, Awe and Ness (Fig. 6.3d), but not in Lochs Morar or Shiel. It was particularly abundant in Loch Lomond with a maximum individual sample density of 3139 individuals per m^3 and a mean annual density of 270 per m^3. The numbers in both the other lochs reached densities of over 600 individuals per m^3 on occasions. The seasonal cycle appeared similar in each loch with moderate numbers during the winter which dropped to very low levels in spring. Numbers started to rise rapidly in May to reach maxima at different times in each loch – June in Loch Lomond, August in Loch Awe and October in Loch Ness.

Bosmina coregoni was found in all five of the lochs investigated (Fig. 6.4a). It was particularly abundant in Lochs Awe and Shiel with maximum individual sample densities of 1076 and 712 individuals per m^3 and mean annual densities of 134 and 203 per m^3 respectively. However it was present in all the lochs in some numbers at all times of the year. A similar seasonal cycle occurred in all lochs with low winter numbers dropping slightly in the spring before starting to rise in May and June. Peak densities occurred in all lochs in June (Lomond and Awe) or July (Ness, Morar and Shiel). Numbers continued

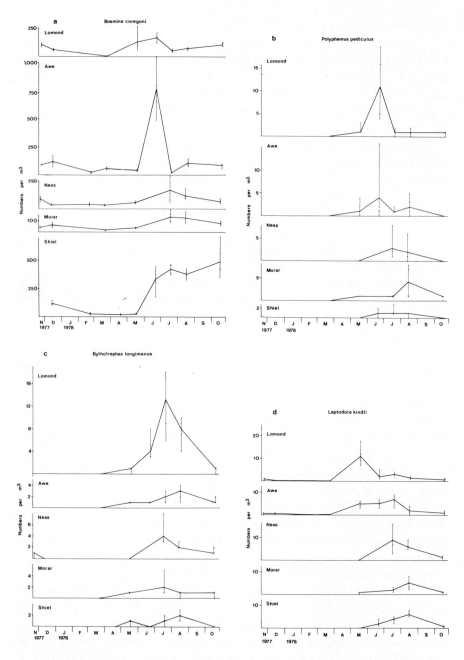

Fig. 6.4 The seasonal abundance in the large lochs from November 1977 to October 1978 of: (a) *Bosmina coregoni*, (b) *Polyphemus pediculus*, (c) *Bythotrephes longimanus*, and (d) *Leptodora kindti*. The values associated with each mean are the actual densities in each haul.

146

to drop in Lochs Ness and Morar but rose again in Lochs Lomond and Awe and especially in Loch Shiel where they reached a seasonal maximum in October.

Polyphemus pediculus occurred in low numbers in the zooplankton of all five lochs (Fig. 6.4b). It was absent during the winter and spring but started to appear in May to reach maximum densities (20 and 16 individuals per m^3 in Lochs Lomond and Awe respectively) in June (Lomond and Awe), July (Ness and Shiel) or August (Morar). Numbers declined thereafter.

Bythothrephes longimanus also occurred in low numbers in all lochs studied (Fig. 6.4c). It was most abundant in Loch Lomond with a maximum individual sample density of 18 individuals per m^3 and a mean annual density of 3 per m^3. This species was absent from the plankton of all lochs during the winter months but started to appear in May and reached maximum numbers in July (Lomond, Ness, Morar) or August (Awe and Shiel). Numbers declined after August in all lochs.

Leptodora kindti, like the two other predatory Cladocera just discussed, was found in the plankton of all five lochs. The densities were always low with maximum individual sample densities in Loch Lomond (18 individuals per m^3). It was absent from the plankton of all of them during the winter (Fig. 6.4d), but started to appear about March or April to reach maxima in May (Lomond), July (Awe and Ness) or August (Morar and Shiel). The populations declined thereafter in all lochs.

Diaptomus gracilis was a common member of the zooplankton of all five lochs at all times of the year (Fig. 6.5a) and in most of them dominated the plankton by number. It was particularly abundant in Lochs Lomond and Shiel with maximum individual sample densities of 2127 and 2697 organisms per m^3 and mean annual densities of 613 and 1064 per m^3 respectively. The seasonal cycle appeared broadly similar in all lochs, with moderate numbers during the winter dropping slightly in spring before rising about April or May to reach maxima in June (Lomond), July (Morar and Shiel) or October (Awe and Ness). Double peaks of density are evident in most of the lochs. Ovigerous females occurred at most times of the year in some of the lochs.

Diaptomus laciniatus occurred only in Loch Shiel (Fig. 6.5c) where it reached a maximum individual sample density of 184 individuals per m^3 with a mean annual density of 28 per m^3. This species was absent from the plankton during the winter and spring, but appeared after May to reach a peak in July. Numbers dropped sharply thereafter to very low levels in October. Ovigerous females occurred from June to October.

Cyclops abyssorum was a common member of the plankton of all five lochs (Fig. 6.5b). It was particularly abundant in Loch Awe and Ness where it

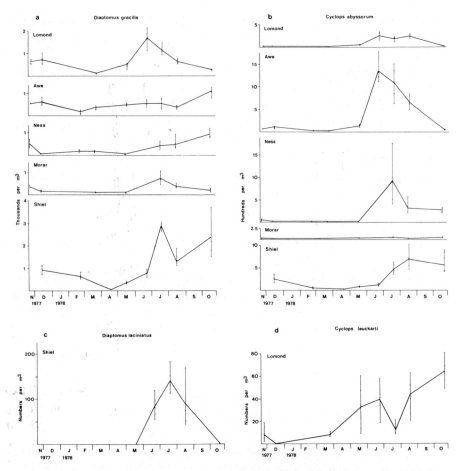

Fig. 6.5 The seasonal abundance in the large lochs from November 1977 to October 1978 of: (a) *Diaptomus gracilis*, (b) *Cyclops abyssorum*, (c) *Diaptomus laciniatus* and (d) *Cyclops leuckarti*. The values associated with each mean are the actual densities in each haul.

reached maximum individual sample densities of 1756 and 1730 organisms per m^3 with mean annual densities of 316 and 192 per m^3 respectively. In all five lochs low numbers during the winter dropped slightly in the spring but started rising in May to reach peak densities in June (Lomond and Awe), July (Ness and Morar) or August (Shiel). Numbers declined thereafter. Ovigerous females occurred at most times of the year in some of the lochs.

Cyclops leuckarti occurred only in Loch Lomond (Fig. 6.5d) where it reached a maximum individual sample density of 80 organisms per m^3 with a mean annual density of 24 per m^3. It occurred in the plankton throughout the

148

year, though numbers were very low during winter. The density increased during the spring to reach a peak in June with a second, higher, peak in October. Ovigerous females occurred from May to August.

Discussion

This study has revealed a number of features of ecological interest concerning the zooplankton of these important large lochs. The results agree broadly with the few previously published records which are available, with some notable exceptions. Only purely open water species are considered. Murray & Pullar (1910) recorded *Diaptomus laciniatus* from Loch Lomond, but this has not been found by subsequent workers (e.g Chapman 1965) and only occurred in Loch Shiel during the present study. Similarly Tippett (1978) found *Holopedium gibberum* in Loch Awe in 1975 and though none was found there in 1977–78 it was found in 1979. Scott (1899) and Scourfield (1908) recorded *Diaptomus laticeps* as being common in Loch Ness, but again none was identified during this study. Such differences are more likely to be due to real changes which have taken place in the zooplankton communities – as have been recorded at Loch Leven (Johnson & Walker 1974) – than to inadequate sampling.

It is often suggested that zooplankton, because of the ubiquity and cosmopolitanism of so many species, are of relatively little use in characterising freshwater bodies. The results of the present study do not agree with this, and it can be seen that as far as these five lochs are concerned a single vertical haul taken in summer will serve to identify the water concerned. If the precedences of the five lochs for various features of the zooplankton are considered (Table 6.6) it can be seen that Loch Lomond is at one end of the

Table 6.6 Precedence for various features of the crustacean zooplankton populations of Lochs Lomond, Awe, Ness, Morar and Shiel

Feature	Lomond	Awe	Ness	Morar	Shiel
Variety of species	1	4	2	5	2
Total zooplankton numbers	3	2	4	5	1
Means of common species	2	1	3	5	3
Maxima of common species	1	2	3	5	4
Means of all species	1	2	4	5	3
Maxima of all species	1	2	3	5	4
First peaks of abundance	1	2	3	4	4
Totals	10	15	22	34	21

list in terms of most features related to the diversity and abundance of the crustacean zooplankton. Loch Morar is just as clearly at the opposite end of the list for the same features, with the other three lochs somewhere in between. These results fit generally with what is known about the catchments of these lochs (Maitland 1981) and their physics (Smith *et al.* 1981b) and chemistry (Bailey-Watts & Duncan 1981a) – see Table 6.1. The subject of plankton communities in relation to lake characteristics is also discussed by Steleanu (1954) and Sprules (1975).

In order to describe the association between different species in communities, Fager (1957) has suggested an index which relates the probabilities of the joint occurrence of the two species to the sum of individual occurrences. This index does not consider negative associations and is a useful preliminary step to the delimitation of communities. An association analysis of the 11 common species of crustacean zooplankton occurring in the five large lochs has been carried out and the results are indicated diagrammatically in Fig. 6.6. There, each species is linked to that with which it is most highly associated, the relevant values being given between the species concerned. As might be expected the six species common to all lochs show 100% association while *Diaphanosoma brachyurum* and *Holopedium gibberum* are linked with these six. *Diaptomus laciniatus* is associated most strongly with both these last two Cladocera, while *Daphnia hyalina* is linked with *Diaphanasoma brachyurum*. *Cyclops leuckarti* is in turn most strongly associated with *Daphnia hyalina*. These results tend to emphasise the uniqueness of Loch Lomond and of Loch Shiel compared to the other three lochs (cf. Fig. 6.2).

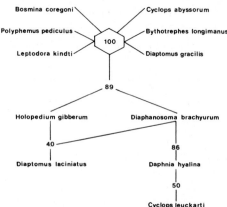

Fig. 6.6 Association analysis of crustacean zooplankton species in the large lochs. Each species is linked to that with which it is most highly associated (Fager 1957), the relevant values being indicated between them as percentages.

150

The pressures imposed by man on the five lochs have been considered in some detail and tabulated by Maitland (1981). There is no evidence to suggest that the zooplankton communities of those under greatest apparent threat (Lomond, Awe and Ness) have deteriorated as a result, compared to those with fewest pressures (Morar and Shiel). Rather is the reverse the case, as is shown in the preceding paragraphs. In particular, the pumped-storage power stations operating at Cruachan on Loch Awe (since 1965) and at Foyers on Loch Ness (since 1975) appear to have had negligible over-all effects on the zooplankton there compared to the other three lochs without pumped-storage schemes.

It has been suggested by Stenson (1972), Nilsson & Pejler (1973), and others that the fish species present in any loch can have a considerable effect on the composition of the zooplankton there. Naturally, the occurrence of zooplankton feeders is of importance in this context. The fish populations of Lochs Lomond, Awe, Ness, Morar and Shiel have been discussed by Maitland *et al.* (1981b). Lochs Awe, Ness and Morar all have populations of charr *(Salvelinus alpinus)* which is predominantly a plankton feeder (Maitland *et al.* 1981a) in the presence (as in these waters) of trout *(Salmo trutta)*. Though Loch Lomond does not have charr, it does have a plankton eating equivalent – the powan *(Coregonus lavaretus)*, found in Scotland only here and in Loch Eck. Loch Shiel has neither powan, nor (surprisingly) does it appear to have charr (Campbell 1979) and the plankton feeding niche here may be vacant. The three-spined stickleback *(Gasterosteus aculeatus)* taken in a plankton haul here is therefore of some interest in this context. It is difficult to speculate on differences in the zooplankton in relation to the fish present in the five large lochs, but it is worth noting that Loch Shiel had the highest mean numbers of zooplankton and was the only loch to contain the large and conspicuous copepod *Diaptomus laciniatus*.

The present results agree generally with the few other ecological studies of zooplankton communities (e.g. Gurney 1923, Smyly 1958) and general life histories (e.g. Colebrook 1956) which have been carried out in Great Britain to date. *Holopedium gibberum* is found in both small and large bodies of water and has a characteristic period of activity and reproduction during the summer months (Hamilton 1958). *Daphnia hyalina* occurred all the year round in three lochs (Lomond, Awe and Ness): in all of these lochs the varieties *galeata* and *lacustris* were found, but normally in different proportions in each loch. The short summer season of *Diaptomus laciniatus* observed in Loch Shiel agrees with the notes by Murray & Pullar (1910) that it normally occurs only from June to October. The data for *Diaptomus gracilis* agree with the more detailed studies of Thomas (1961) and Chapman (1969).

151

Though *Cyclops leuckarti* (only occurring in Loch Lomond in the present study) is absent from the winter plankton of some English lakes (Smyly 1961), it is found all the year round in Loch Lomond (Chapman 1972; Fig. 6.5d).

Acknowledgements

This research was carried out under contract to the North of Scotland Hydro-Electric Board and the Nature Conservancy Council. We are grateful to Mrs P. Duncan and Mr A. J. Rosie who undertook most of the field sampling, often in difficult and arduous conditions. Some of the figures were drawn by Mrs S. M. Adair and the manuscript was typed by Mrs M. S. Wilson. Useful comments on early drafts of the paper were received from Dr A. E. Bailey-Watts and Mr D. H. Jones.

References

Bailey-Watts, A. E. & Duncan, P., 1981a. The ecology of Scotland's largest lochs: Lomond, Awe, Ness, Morar and Shiel. Ed. P. S. Maitland. Chemical characterisation – a one-year comparative study. Chap. 3. This volume.

Bailey-Watts, A. E. & Duncan, P., 1981b. The ecology of Scotland's largest lochs: Lomond, Awe, Ness, Morar and Shiel. Ed. P. S. Maitland. The phytoplankton. Chap. 4. This volume.

Bailey-Watts, A. E. & Duncan, P., 1981c. The ecology of Scotland's largest lochs: Lomond, Awe, Ness, Morar and Shiel. Ed. P. S. Maitland. A review of macrophyte studies. Chap. 5. This volume.

Campbell, R. N., 1979. Ferox trout, *Salmo trutta* L., and charr, *Salvelinus alpinus* (L.)., in Scottish lochs. J. Fish. Biol. 14: 1–29.

Chapman, M. A., 1965. Ecological studies on the zooplankton of Loch Lomond. Ph. D. Thesis, University of Glasgow.

Chapman, M. A., 1969. The bionomics of *Diaptomus gracilis* Sars (Copepoda, Calanoida) in Loch Lomond, Scotland. J. Anim. Ecol. 38: 257–284.

Chapman, M. A., 1972. The annual cycles of the limnetic cyclopoid Copepoda of Loch Lomond, Scotland. Int. Revue ges. Hydrobiol. Hydrogr. 57: 895–911.

Colebrook, J. M., 1956. The seasonal life cycles of some of the planktonic Crustacea of Windermere. Unpublished Manuscript, Freshwater Biological Association, Ambleside.

Fager, E. W., 1957. Determination and analysis of recurrent groups. Ecology, 38: 586–595.

George, D. G. & Jones, D. H., 1981. Spatial studies of the zooplankton of Scotland's largest lochs: Lomond, Awe, Ness, Morar and Shiel. In preparation.

Goldspink, C. R. & Scott, D. B. C., 1971. Vertical migration of *Chaoborus flavicans* in a Scottish loch. Freshwat. Biol. 1: 411–421.

Gurney, R., 1923. The crustacean plankton of the English Lake District. J. Linn. Soc. Zool. 35: 412–447.

Hamilton, J. D., 1958. On the biology of *Holopedium gibberum* Zaddach (Crustacea: Cladocera). Verh. int. Verein theor. angew. Limnol. 13: 785–788.

Johnson, D. & Walker, A. F., 1974. The zooplankton of Loch Leven, Kinross. Proc. R. Soc. Edinb. 74: 285–294.

Lamond, H., 1931. Loch Lomond. Glasgow: Jackson. 340 pp.

McGowan, J. A. & Fraundorf, V. J., 1966. The relationship between size of net used and estimates of zooplankton diversity. Limnol. Oceanogr. 11: 456–469.

Maitland, P. S., 1969. The reproduction and fecundity of the powan, *Coregonus clupeoides* in Loch Lomond, Scotland. Proc. R. Soc. Edinb. 70: 233–264.

Maitland, P. S., 1981. The ecology of Scotland's largest lochs: Lomond, Awe, Ness, Morar and Shiel. Ed. P. S. Maitland. Introduction and catchment analysis. Chap. 1. This volume.

Maitland, P. S., Greer, R. B., Campbell, R. N. & Friend, G. F., 1981a. The status and biology of arctic charr, *Savelinus alpinus* (L) in Scotland. In preparation.

Maitland, P. S., Smith, B. D. & Adair, S. M., 1981b. The ecology of Scotland's largest lochs: Lomond, Awe, Ness, Morar and Shiel. Ed. P. S. Maitland. The fish and fisheries. Chap. 9. This volume.

Maitland, P. S., Smith, I. R., Bailey-Watts, A. E., George, D. G., Lyle, A. A., Smith, B. D., Duncan, P., Rosie, A. J., Dennis, G. M. & Carr, M. J., 1981c. Twenty-four hour studies of the effects of pumped-storage power stations on water and plankton in Loch Awe (Cruachan) and Loch Ness (Foyers). In preparation.

Murray, J., 1904. Notes on the biology of Loch Morar. Geogrl. J. 24: 77–79.

Murray, J., 1905. On the distribution of the pelagic organisms in Scottish lakes. Proc. R. Phys. Soc. Edinb. 16: 51.

Murray, J. & Pullar, L., 1910. Bathymetrical survey of the freshwater lochs of Scotland. Edinburgh: Challenger. Vols. 1–6.

Nilsson, N. A. & Pejler, B., 1973. On the relation between fish fauna and zooplankton composition in north Swedish lakes. Rep. Inst. Freshwat. Res. Drottningholm. 53: 51–77.

North of Scotland Hydro-Electric Board. 1976. Report on Craigroyston pumped storage. Edinburgh: Unpublished report.

Scott, T., 1899. The invertebrate fauna of the inland waters of Scotland – report on special investigation. Rep. Fishery Bd. Scot. 17: 133–204.

Scott, T., 1901. Land, freshwater and marine crustacea. Brit. Ass. Handb. Glasg. 328–358.

Scourfield, D. J., 1908. The biological work of the Scottish lake survey. Int. Revue ges. Hydrobiol. Hydrogr. 1: 177–192.

Slack, H. D., 1957. Studies on Loch Lomond. I. Glasgow: Blackie.

Slack, H. D., 1965. The profundal fauna of Loch Lomond. Proc. R. Soc. Edinb. 69: 272–297.

Smith, B. D., Maitland, P. S., Young, M. R. & Carr, M. J., 1981a. The ecology of Scotland's largest lochs: Lomond, Awe, Ness, Morar and Shiel. Ed. P. S. Maitland. The littoral zoobenthos. Chap. 7. This volume.

Smith, I. R., Lyle, A. A. & Rosie, A. J., 1981b. The ecology of Scotland's largest lochs: Lomond, Awe, Ness, Morar and Shiel. Ed. P. S. Maitland. Comparative physical limnology. Chap 2. This volume.

Smyly, W. J. P., 1958. The Cladocera and Copepoda (Crustacea) of the tarns of the English Lake District. J. Anim. Ecol. 2: 87–103.

Smyly, W. J. P., 1961. The life cycle of the freshwater copepod *Cyclops leuckarti* Claus in Esthwaite Water. J. Anim. Ecol. 30: 153–171.

Sprules, W. M., 1975. Factors affecting the structure of limnetic crustacean zooplankton communities in several central Ontario lakes. Verh. int. Verein. Limnol. theor. angew. 19: 635–643.

Steleanu, A., 1954. The problem of the effect on and control of lake plankton by physical factors. Acta Hydrophys. 1: 45–56.

Stenson, J. A. E., 1972, Fish predation effects on the species composition of the zooplankton community in eight small forest lakes. Rep. Inst. Freshwat. Res. Drottningholm. 52: 132–148.

Thomas, M. P., 1961. Some factors influencing the life history of *Diaptomus gracilis* Sars. Verh. int. Verein. theor. angew. Limnol. 14: 943–945.

Tippett, R., 1978. The effect of the Cruachan pumped-storage hydro-electric scheme on the limnology of Loch Awe. University of Glasgow: Unpublished Report.

Unesco, 1968. Zooplankton sampling. Monographs on Oceanographic Methodology. Geneva: UNESCO.

7. The littoral zoobenthos

B. D. Smith, P. S. Maitland, M. R. Young & M. J. Carr

Abstract

The littoral substrate and its associated zoobenthos is described from a number of samples collected around the shorelines of these five lochs, during autumn 1977. Further samples were taken from autumn 1977 to autumn 1978 to provide comparative information on species diversity, seasonality, growth, life-cycles and depth distribution of littoral zoobenthos. Loch Lomond had the most diverse and abundant fauna while those of Lochs Shiel and Morar were very impoverished in both respects. Lochs Lomond, Awe and Ness had many features in common but Loch Morar and Shiel were distinct from these three lochs and from each other. The similarities and differences are discussed in relation to environmental factors, and the ecology of various species of ecological interest is also considered. Man's effect on the environment is reviewed, particularly in relation to hydro-electric schemes.

Introduction

In spite of their importance, both as a natural resource and ecologically, the largest lochs of Scotland have been surprisingly neglected by limnologists. Increasing pressures from Man have highlighted our ignorance of these systems and emphasized the need for comparative baseline studies before major changes take place. During 1977, 1978 and 1979 a survey was made of the littoral zoobenthos of five of Scotland's largest lochs – Lomond, Awe, Ness, Morar and Shiel.

The lochs are situated in the north and west Highlands of Scotland and are all large oligotrophic water bodies, with the exception perhaps of Loch Lomond with part of its south basin and catchment lying to the south of the

Table 7.1 Basic physical and chemical characteristics of relevance to the littoral zoobenthos study of Lochs Lomond, Awe, Ness, Morar and Shiel

Feature	Lomond	Awe	Ness	Morar	Shiel
Catchment area (km²)	781	840	1775	168	248
Length of shore-line					
(excluding islands) (km)	103	114	85	51	78
Mean shore-line gradient					
(m/m)	18.99	14.09	3.71	9.92	15.97
Annual mean water level					
(m above sea level)	7.93	36.21	15.80	10.13	4.54
Catchment (% forest)	11.0	19.6	14.4	3.2	15.7
Catchment (% arable)	17.0	1.6	1.9	0.6	1.4
Catchment					
(population numbers)	12 218	777	3128	162	354
* pH	6.8	6.9	6.7	6.6	6.1
* Alkalinity (as $CaCO_3$ mg l⁻¹)	6.4	9.0	4.5	3.3	2.1
* Ca (mg l⁻¹)	3.0	4.0	2.0	1.2	1.0
* Conductivity (K_{20} μS cm⁻¹)	34	41	30	35	29
* Total organic Carbon (mg l⁻¹)	2.1	3.3	2.9	1.6	1.7
* $NO_3 + NO_2$ N (μg l⁻¹)	200	100	110	75	55

* Mean values for the study period.

Highland Boundary Fault. A comparison of some major features of their morphometry and chemistry is given in Table 7.1. Further information on the morphometry is available in Smith *et al.* (1981), on the catchment in Maitland (1981) and on the chemistry in (Bailey-Watts & DUncan 1981).

Lochs Morar and Shiel are rather isolated and largely untouched by Man although Loch Morar has a small conventional hydro-electric scheme operational since 1948, on its outflow – the River Morar. In contrast Lochs Awe and Ness are part of existing pumped-storage schemes operational since 1965 and 1975 respectively. Further, Loch Ness forms part of the Caledonian Canal system which was opened in 1822. Loch Lomond has several agencies using its water resources. Water abstraction from the southern basin began in 1971 and a barrage was erected at the outflow, the River Leven, to control the water level. The U.K.'s largest conventional hydro-electric station (Loch Sloy) has been discharging into the northern basin since 1950. There is also particularly large tourism pressure on this loch (which lies within the industrial central belt of Scotland). Further there is a proposal by the North of Scotland Hydro-Electric Board (1976) to build a large pumped-storage power scheme beside the north basin of the loch.

There is very little information on the zoobenthos of these lochs apart from

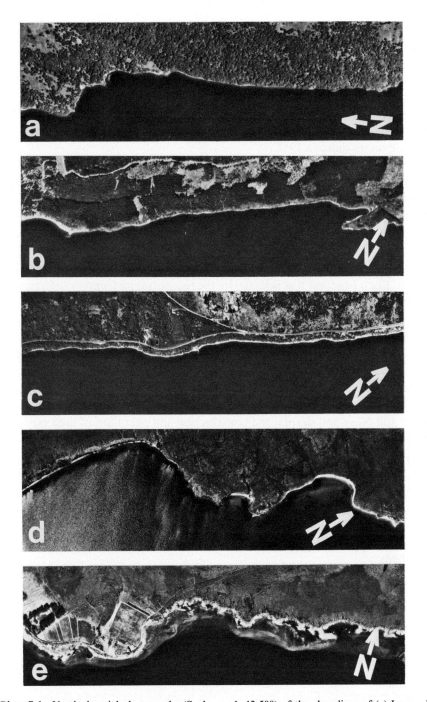

Plate 7.1 Vertical aerial photographs (Scale *ca.* 1 : 12 500) of the shorelines of (a) Lomond (b) Awe (c) Ness (d) Morar (e) Shiel. (Photos: Crown Copyright Reserved).

157

Loch Lomond. A brief note on the Coleoptera from Loch Awe (Chitty, 1892, 1893) and some data on Lochs Ness and Morar resulting from the bathymetrical survey of Scottish lochs (Scourfield 1908, Murray 1910) and a special investigation of Scottish fresh waters by Scott (1899) appear to be the extent of the early literature. In contrast, Loch Lomond has been the subject of biological studies for many years e.g. Fergusson (1901, 1910), Scott (1899), etc. and is a research site for the University of Glasgow which set up a field station on its shore in 1947.

General works on the fauna of Loch Lomond include the studies of Slack (1957) and Weerekoon (1956a, b). A wide range of invertebrate animals inhabiting Loch Lomond has been the subject of more specific studies e.g. Gastropoda (Hunter, 1953a, b, c, 1954, 1957, 1961a, b, Hunter *et al.* 1964), Bivalvia (Hunter & Slack 1958), Ephemeroptera (Calow 1974), Plecoptera (Sinclair 1953) and Diptera (Weerekoon 1953, Lawson 1957). Some of the influences of Man on this environment have also been examined (Maitland 1972). Loch Lomond was selected by the recent Nature Conservation Review (Ratcliffe 1977) as a key site in the selection of sites of particular scientific merit.

Lochs Morar and Shiel were also included in this review. Loch Morar is the most extreme large oligotrophic lake in Britain and is ranked as a Grade 1 site, while Loch Shiel is a Grade 2 alternative (for conservation) to Loch Morar. The review includes a limited amount of information on the littoral zoobenthos of these two lochs. Despite the size and importance of Lochs Awe and Ness and despite the construction and operation of their pumped-storage schemes, there appears to be no information about the zoobenthos of either loch.

The aims of the present study were firstly to obtain a comprehensive species list for each loch and for each substrate type within the loch; secondly to identify seasonal differences in the fauna and thirdly to study the depth preferences of littoral zoobenthos.

Methods

In order to define the main shore substrate types of each loch, the littoral zone was systematically surveyed. The perimeter of each loch was navigated by small boat and the substrate was classified subjectively, according to the Wentworth Scale (Wentworth 1922). A note was made of the nature of bank vegetation and local land use. (Examples of stretches of shore-line are shown in Plate 7.1). In order to provide a comprehensive species list and an account of the communities associated with the main substrate types (Plate 7.2), a

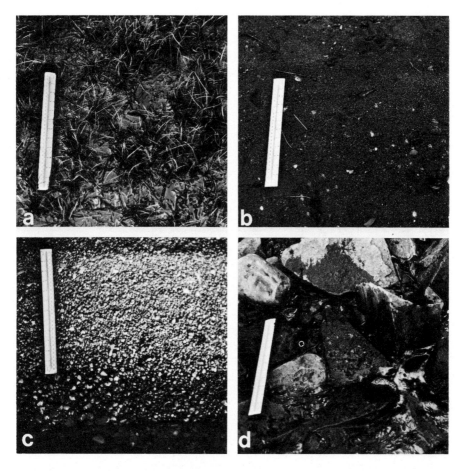

Plate 7.2 Composite photograph of the main substrate types (scale indicated by 30 cm rule) found in the large lochs (a) silt (b) sand (c) gravel (d) stones and boulders. (Photos: B. D. Smith, P. S. Maitland.)

large number of sites around each loch (to include all substrate types) were selected (Fig. 7.1a-e). Samples were collected from Lochs Morar and Shiel by MRY in late September and early October 1977; those from Lochs Lomond, Awe and Ness by BDS and PSM in November and December 1977.

These benthic invertebrate samples were obtained by timed collection with a 27 cm diameter circular hand net (mesh size 0.5 mm), in the manner described by Macan (1957) and Hynes (1961). In most cases samples were collected for 10 min at depths from 0-50 cm. Under these circumstances samples consisted of two sub-samples of 3 min stone-kicking and 2 min stone-washing. Where the substrate was unsuitable for such methods 2 sub-

160

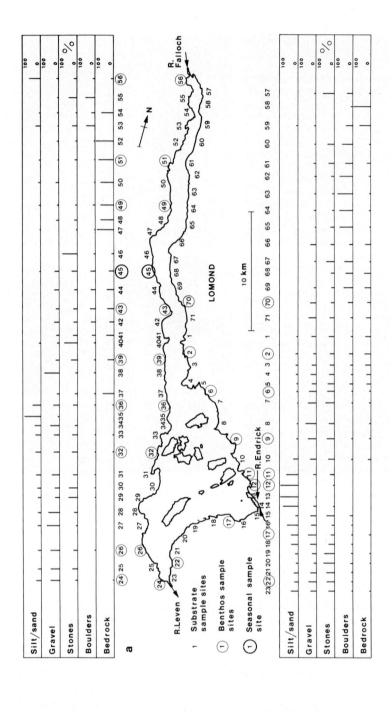

Fig. 7.1a The distribution of sample sites around the perimeter of Loch Lomond (the site is marked at the centre of each section surveyed, approx 0.5–2.0 km long). Also shown is the percentage composition of the substrate of each site (bedrock includes man-made material, e.g. concrete embankments).

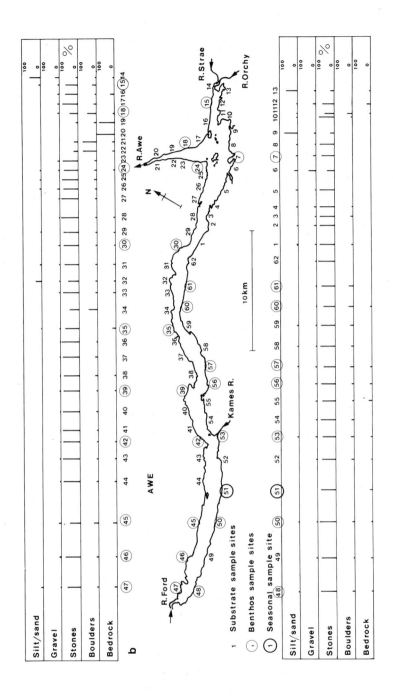

Fig. 7.1b The distribution of sample sites around the perimeter of Loch Awe (the site is marked at the centre of each section surveyed, approx. 0.5–3.5 km long). Also shown is the percentage composition of the substrate of each site (bedrock includes man-made material).

161

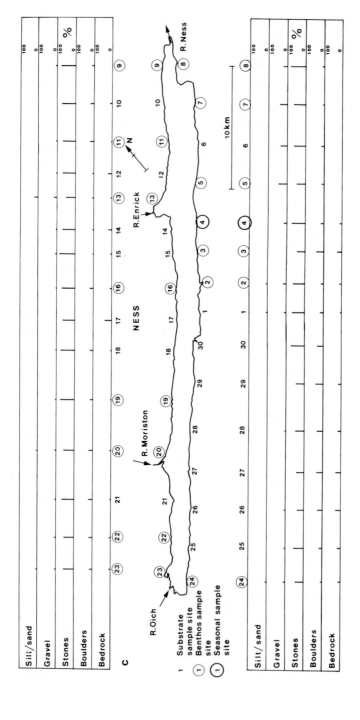

Fig. 7.1c The distribution of sample sites around the perimeter of Loch Ness (the site is marked at the centre of each section surveyed approx 1.5–3.0 km long). Also shown is the percentage composition of the substrate of each site (bedrock includes man-made material).

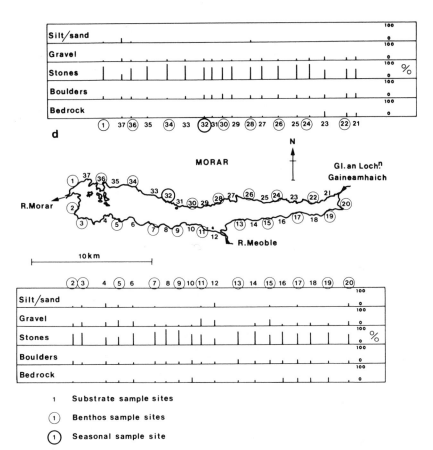

Fig. 7.1d The distribution of sample sites around the perimeter of Loch Morar (the site is marked at the centre of each section surveyed, approx 1.0-1.5 km long). Also shown is the percentage composition of the substrate of each site.

samples of 3 min stone-kicking alone (e.g. in gravel) or 2 min 'netting' (e.g. in reeds beds) were used.

Following this shoreline survey a typical and apparently very comparable stony littoral shore was selected on each loch (Fig. 7.1a-e; Plates 7.3 a-e). Seasonal sampling was carried out from December 1977 to October 1978 on 9 occasions at Loch Lomond and on 8 occasions at the remaining 4 lochs. The hand net described above was used to collect the samples which were taken at depths of 10, 20, 30, 40, 50 cm at each site. The duration of the sample was varied according to the abundance of the fauna, but consisted basically of a 2 min wash and a 1 min kick at each depth.

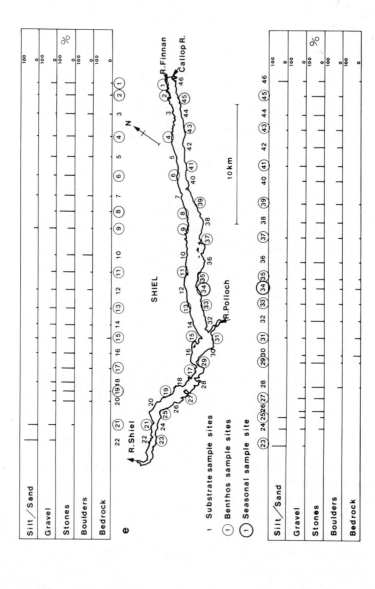

Fig. 7.1e The distribution of sample sites around the perimeter of Loch Shiel (the site is marked at the centre of each section surveyed, approx. 1.0-2.0 km long). Also shown is the percentage composition of the substrate of each site.

Plate 7.3a Seasonal sampling site at Loch Lomond looking north, showing dense bank cover of deciduous trees. (Photo: B. D. Smith.)

165

Plate 7.3b Seasonal sampling site at Loch Awe looking south, showing dense bank cover of deciduous trees. (Photo: B. D. Smith.)

Plate 7.3c Seasonal sampling site at Loch Ness looking north, showing dense bank cover of deciduous trees. (Photo: B. D. Smith.)

Plate 7.3d Seasonal sampling site at Loch Morar looking east, showing bank cover of bracken and heather. Few trees occur around Loch Morar. (Photo: K. H. Morris.)

Plate 7.3e   Seasonal sampling site at Loch Shiel looking south showing bank cover of deciduous trees. (Photo: P. S. Maitland.)

All samples were labelled and preserved in the field with 4% formalin. The material was sieved (mesh size: 0.5 mm) and hand sorted in the laboratory where the invertebrate animals collected were identified as far as possible. Selected species (*Lymnaea peregra, Ameletus inopinatus, Ephemerella ignita, Diura bicaudata* and *Chloroperla torrentium*) which were reasonably abundant and occurred at more than one loch, were measured regularly for total length. The mayfly and stonefly larvae were measured from the tip of the head to the base of the cerci. Additional *L. peregra* were collected from each loch and measured from the base of the aperture to the top of the spire.

Description of the lochs and their substrates

General

Loch Lomond has by far the greatest surface area and the most gently sloping shore. In contrast, Loch Ness, which is the largest loch by volume, has very steeply sloping shores. Lochs Morar and Shiel are much smaller and Loch Morar has a similarly steeply sloping shore to Loch Ness. The gradients of the shores of Lochs Shiel and Awe are similar and fall between these extremes.

Lochs Lomond, Awe and Ness have large catchments and similar water chemistry. However, Loch Awe has a relatively high alkalinity, conductivity and total organic carbon content and the dissolved nitrogen content of Loch Lomond is double that of Loch Awe (which presumably reflects the higher population of the Lomond catchment). Lochs Morar and Shiel have considerably smaller catchments than the preceeding three lochs. The pH of both lochs is slightly lower and they are chemically much poorer than Lochs Lomond, Awe and Ness. Of the two, Loch Morar has the slightly higher conductivity (probably related to its close proximity to the sea) and alkalinity while Loch Shiel has a slightly higher total organic carbon.

Littoral substrates

The bank cover of all the lochs is largely grass and herbs (see Table 7.2). Lochs Lomond, Awe and Ness have a high percentage of trees along their shores while Loch Shiel has approximately half the tree cover of these three lochs. Loch Morar is largely without tree cover along the shore (Plates 7.3a-e).

Stones form the main components of the littoral substrate of all the lochs.

170

Table 7.2 The percentage composition of littoral substrate and bank cover of Lochs Lomond, Awe, Ness, Morar and Shiel

Feature	Lomond	Awe	Ness	Morar	Shiel
Tree cover	71	72	79	17	37
Bank cover					
Sand					
(0.063 mm–2.00 mm)	1	1	–	2	3
Gravel					
(2.00 mm–6.40 cm)	7	1	–	11	16
Stones					
(6.40 cm–25.60 cm)	1	–	–	–	–
Boulders					
(> 25.60 cm)	2	–	4	–	–
Bedrock	11	3	9	7	6
Grass	38	66	50	46	54
Herbs	31	22	36	33	21
Man-made	9	7	1	1	–
Overall substrate					
Silt					
(0–0.063 mm)	1	1	1	1	2
Sand					
(0.063 mm–2.00 mm)	15	5	1	4	9
Gravel					
(2.00 mm–6.40 cm)	22	6	9	13	20
Stones					
(6.40 cm–25.60 cm)	25	69	63	62	47
Boulders					
(> 25.60 cm)	16	12	16	13	16
Bedrock	16	4	10	8	6
Man-made	5	3	–	–	–

Lochs Awe, Ness and Morar have more than 60% stony shores with only small areas of silt, sand and gravel. Loch Shiel has a slightly higher proportion of silt, sand and gravel while Loch Lomond has extensive beaches of material of this smaller size range.

The distribution of these substrates is included in Fig. 7.1a-e. The silt, sand and gravelly shores of Loch Lomond are mainly located in the southern basin particularly around the River Endrick inflow. (There is, however, a small area of silt at the northernmost end of the loch where the River Falloch enters). The silty/sandy areas in Loch Awe are also located by large river inflows – in the south west, the Rivers Kames and Ford, and in the east, the

Rivers Orchy and Strae. Again, in Loch Ness, deposits of this smaller substrate class are found only around river inflows, especially the Enrick, Moriston and Oich. The largest areas of sand and gravel in Loch Shiel tend to be restricted to the western end of the loch (Plate 7.1e) but there are small areas of silt and sand at the inflows of the Rivers Finnan and Callop. There are only very small areas (< 5%) sand and silt around Loch Morar (shown near the outflow in Plate 7.1d).

Results

Species diversity

The results of the zoobenthos sampling from the shore-line survey of the

Table 7.3   The taxa collected during the shoreline survey of Lochs Lomond, Awe, Ness, Morar and Shiel (% of total and mean number per 10 min collection)

Taxa	Lomond		Awe		Ness		Morar		Shiel	
	%	mean	%	mean	%	mean	%	mean	%	mean
COELENTERATA	0	0	0	1	0	0	0	0	0	0
Tricladida	1	19	0	8	1	7	0	0	0	0
NEMATODA	0	1	0	0	0	1	0	0	1	1
Gastropoda	2	41	1	10	1	11	0	0	0	0
Bivalvia	1	11	0	0	0	0	0	0	1	1
Oligochaeta	29	604	25	445	22	345	45	32	60	62
Hirudinea	0	8	0	1	0	1	0	0	0	0
Hydracarina	0	1	0	0	0	0	0	0	0	0
Cladocera	0	5	1	27	0	4	0	0	0	0
Ostracoda	0	0	0	1	0	0	0	0	0	0
Copepoda	0	2	0	0	0	0	0	0	0	0
Malacostraca	8	170	4	76	3	44	0	0	0	0
Collembola	0	0	0	0	0	1	0	0	0	0
Ephemeroptera	41	860	37	649	18	284	12	9	5	5
Plecoptera	6	123	4	69	30	463	31	22	20	21
Odonata	0	1	0	0	0	0	0	0	0	0
Hemiptera	3	65	3	47	7	115	0	0	1	1
Coleoptera	2	42	1	19	2	28	7	5	2	2
Megaloptera	0	0	0	0	0	1	0	0	0	0
Trichoptera	4	95	9	165	7	107	1	1	2	2
Diptera	3	53	15	258	9	137	4	3	8	9
Total	100	2101	100	1776	100	1549	100	72	100	104

Table 7.4 Littoral zoobenthos identified to species (mean number per 10 min collection) from the shoreline survey of Lochs Lomond, Awe, Ness, Morar and Shiel

Species	Lomond	Awe	Ness	Morar	Shiel
Polycelis felina (Dalyell)	0	0	1	0	0
Planaria torva (Muller)	6	0	0	0	0
Phagocata woodworthi Hyman	0	0	6	0	0
Dendrocoelum lacteum (Muller)	2	0	0	0	0
Potamopyrgus jenkinsi (Smith)	3	0	0	0	0
Lymnaea truncatula (Muller)	0	2	0	0	0
L. peregra (Muller)	7	6	11	0	0
Physa fontinalis (L.)	25	0	0	0	0
Planorbis laevis Alder	2	0	0	0	0
P. albus Muller	2	0	0	0	0
Ancylus fluviatilis Muller	2	1	0	0	0
Stylaria lacustris (L.)	159	57	45	0	1
Lumbriculus variegatus (Muller)	103	23	118	5	6
Stylodrilus heringianus Claparede	71	41	11	0	0
Eiseniella tetraedra (Savigny)	5	4	0	0	0
Glossiphonia complanata (L.)	2	0	1	0	0
Helobdella stagnalis (L.)	6	0	0	0	0
Sida crystallina (Muller)	0	2	0	0	0
Eurycercus lamellatus Muller	4	25	3	0	0
Asellus aquaticus (L.)	157	0	36	0	0
A. meridianus Racovitza	0	0	6	0	0
Gammarus lacustris Sars	0	38	0	0	0
G. pulex (L.)	13	33	1	0	0
Ameletus inopinatus Eaton	0	0	2	0	0
Centroptilum luteolum (Muller)	76	13	0	0	0
Cloeon dipterum (L.)	0	0	21	0	0
C. simile Eaton	7	626	188	0	0
Ecdyonurus venosus (Fabricius)	15	0	46	0	0
E. dispar (Curtis)	1	1	0	8	5
Leptophlebia marginata (L.)	10	3	20	0	0
L. vespertina (L.)	7	0	4	0	0
Ephemerella ignita (Poda)	0	0	0	1	0
Ephemera danica Muller	4	1	0	0	0
Caenis moesta Bengtsson	633	0	0	0	0
C. horaria (L.)	88	0	0	0	0
Amphinemura sulcicollis (Stephens)	2	0	65	0	0
Nemoura cinerea (Retzius)	0	9	0	0	0
N. avicularis Morton	19	0	7	0	0
N. cambrica (Stephens)	0	0	28	0	0
Leuctra inermis Kempny	0	0	32	0	0
L. hippopus (Kempny)	1	0	13	0	3
L. fusca (L.)	0	0	0	0	8
Capnia bifrons (Newman)	0	40	23	0	0

173

Table 7.4 (continued)

Species	Lomond	Awe	Ness	Morar	Shiel
C. atra Morton	4	0	0	0	0
Diura bicaudata (L.)	1	14	26	1	0
Chloroperla torrentium (Pictet)	94	5	98	0	8
C. tripunctata (Scopoli)	2	0	14	20	0
Ischnura elegans (Linden)	1	0	0	0	0
Notonecta glauca L.	0	0	1	0	0
Micronecta poweri (Douglas & Scott)	0	2	0	0	0
Cymatia bonsdorffi (Sahlberg)	1	3	0	0	0
Callicorixa praeusta (Fieber)	0	1	1	0	0
Sigara dorsalis (Leach)	11	27	67	0	0
S. distincta (Fieber)	5	8	29	0	0
S. falleni (Fieber)	43	0	0	0	0
S. fossarum (Leach)	1	0	2	0	0
S. scotti (Fieber)	0	6	7	0	0
Haliplus fulvus (Fabricius)	0	1	0	0	0
Deronectes depressus (Fabricius)	2	3	3	0	0
Orectochilus villosus (Muller)	0	1	0	0	0
Esolus parallelepipedus (Muller)	0	0	3	0	0
Limnius volkmari (Panzer)	15	0	2	3	0
Sialis lutaria (L.)	0	0	1	0	0
Agapetus fuscipes Curtis	3	0	2	0	0
Plectrocnemia conspersa (Curtis)	0	1	0	0	0
Polycentropus flavomaculatus (Pictet)	3	2	6	0	0
Cyrnus flavidus McLachlan	1	0	1	0	0
Tinodes waeneri (L.)	5	0	25	0	0
Phryganea varia Fabricius	0	0	6	0	0
Lepidostoma hirtum (Fabricius)	8	0	23	0	1
Sericostoma personatum (Spence)	2	0	1	0	0
Total	1634	999	1006	38	32
Total number of species	47	31	43	6	7

lochs are given in Tables 7.3 and 7.4. Additional species collected during seasonal sampling are given in Table 7.8. Loch Lomond with a total of 148 taxa identified from it, had the most diverse and abundant fauna of the five lochs. Loch Ness had the second most varied fauna (125 taxa) and Loch Awe the third (116 taxa) but the fauna of Loch Awe is the more abundant. Lochs' Shiel (97 taxa) and Morar (71 taxa) faunas fall considerably behind in abundance and diversity.

Certain groups of animals were important constituents of the faunas of all five lochs, notably Oligochaeta, Ephemeroptera, Plecoptera, Coleoptera, Trichoptera and Diptera. Tricladida, Gastropoda, Hirudinea, Cladocera, Malacostraca and Hemiptera are important only in Lochs Lomond, Awe and Ness.

Table 7.4 gives the mean numbers of littoral zoobenthos identified to species which were collected during the shoreline survey. Loch Lomond had 47 such species which occurred frequently enough for a mean of at least 1 individual to be taken per ten min collection. The second most varied fauna was found at Loch Ness, with a total of 43 species, and then Loch Awe with 31 species, Lochs Shiel and Morar had only 7 and 6 species respectively.

Table 7.5 Littoral zoobenthos identified to species (mean number per 10 min collection) from the silty/sandy shores of Lochs Lomond, Awe, Ness and Shiel. Authorities are included for only those species not in Table 7.4

Species	Lomond	Awe	Ness	Shiel
Polycelis felina	0	0	2	0
Crenobia alpina (Dana)	0	0	1	0
Planaria torva	1	0	0	0
Dendrocoelum lacteum	1	0	0	0
Potamopyrgus jenkinsi	1	0	0	0
Lymnaea peregra	2	0	5	0
Physa fontinalis	14	1	0	0
Planorbis laevis	2	0	0	0
Stylaria lacustris	316	145	158	0
Lumbriculus variegatus	147	35	187	3
Stylodrilus heringianus	67	9	8	2
Eiseniella tetraedra	1	1	0	0
Theromyzon tessulatum (Muller)	0	1	0	0
Glossiphonia complanata	1	0	0	0
Helobdella stagnalis	0	0	1	0
Sida crystallina	0	10	0	0
Eurycercus lamellatus	14	123	14	0
Asellus aquaticus	185	0	85	0
A. meridianus	0	0	19	0
Gammarus lacustris	0	8	0	0
G. pulex	0	1	4	0
Siphlonurus lacustris Eaton	1	0	0	0
Baetis rhodani (Pictet)	1	0	0	0
Baetis niger (L.)	0	1	0	0
Centroptilum luteolum	310	50	26	0
Cloeon dipterum	1	0	85	0
C. simile	24	3127	751	2

Table 7.5 (continued)

Species	Lomond	Awe	Ness	Shiel
Leptophlebia marginata	21	1	78	0
L. vespertina	20	0	15	0
Ephemera danica	5	0	0	0
Caenis moesta	500	0	0	0
C. horaria	46	0	0	0
Amphinemura sulcicollis	0	0	1	0
Nemoura cinerea	0	32	0	0
N. avicularis	17	0	139	0
Leuctra hippopus	2	0	1	0
Capnia bifrons	0	0	2	0
Diura bicaudata	0	2	0	0
Chloroperla torrentium	7	0	1	0
Ischnura elegans	1	0	0	0
Notonecta glauca	0	0	4	0
Cymatia bonsdorffi	4	16	0	0
Callicorixa praeusta	0	4	3	0
Sigara dorsalis	35	76	259	2
S. distincta	8	0	115	0
S. falleni	153	39	0	0
S. fossarum	3	0	6	0
S. scotti	0	32	30	2
Haliplus fulvus	0	2	0	0
H. variegatus Sturm	0	1	0	0
Deronectes depressus	1	0	13	0
Oreodytes septentrionalis (Gyllenhal)	1	0	0	0
Coelambus novemlineatus Stephen	0	0	0	2
Orectochilus villosus	0	1	0	0
Limnius volkmari	1	0	0	0
Sialis lutaria	0	0	4	0
Polycentropus flavomaculatus	1	0	10	0
Cyrnus flavidus	2	0	4	0
Tinodes waeneri	3	0	0	0
Phryganea varia	0	0	24	0
Lepidostoma hirtum	10	0	4	0
Sericostoma personatum	1	0	4	0
Total	1931	3718	2063	13
Total number of species	40	24	34	6
Number of samples	5	4	4	3

Table 7.6 Littoral zoobenthos identified to species (mean number per 10 min collection) from the gravel shores of Lochs Lomond, Awe, Ness, Morar and Shiel. Authorities are included only for those species not in Table 7.4

Species	Lomond	Awe	Ness	Morar	Shiel
Planaria torva	6	0	0	0	0
Crenobia alpina	0	0	1	0	0
Phagocata woodworthi	0	0	32	0	0
Dendrocoelum lacteum	3	0	0	0	0
Lymnaea peregra	7	4	2	0	0
Physa fontinalis	41	0	0	0	0
Ancylus fluviatilis	7	0	0	0	0
Stylaria lacustris	53	85	12	0	4
Lumbriculus variegatus	85	30	334	0	0
Stylodrilus heringianus	110	57	29	1	0
Eiseniella tetraedra	6	3	8	0	0
Glossiphonia complanata	0	0	1	0	0
Helobdella stagnalis	7	0	0	0	0
Asellus aquaticus	202	1	80	0	0
A. meridianus	0	0	8	0	0
Gammarus lacustris	0	43	0	0	0
G. pulex	39	52	0	0	0
Ameletus inopinatus	0	0	2	0	0
Centroptilum luteolum	47	1	0	0	0
Ecdyonurus venosus	2	0	101	0	0
E. dispar	5	0	0	2	0
Leptophlebia marginata	9	3	1	0	0
Ephemerella ignita	0	1	0	0	0
Ephemera danica	6	1	0	0	0
Caenis moesta	496	0	0	0	0
C. horaria	80	0	0	0	0
Amphinemura sulcicollis	5	0	171	0	0
Nemoura avicularis	15	0	1	0	0
Leuctra inermis	0	0	56	0	0
L. hippopus	0	0	3	0	0
L. nigra (Oliver)	1	0	0	0	0
Capnia bifrons	0	140	10	0	0
C. atra	2	0	0	0	0
Diura bicaudata	0	21	15	0	0
Chloroperla torrentium	194	7	220	0	0
C. tripunctata	0	0	19	0	0
Micronecta poweri	0	4	0	0	0
Sigara dorsalis	4	0	0	0	4
S. distincta	1	0	0	0	0
S. scotti	0	0	0	0	1
Haliplus fulvus	0	1	0	0	0
Deronectes depressus	7	5	0	0	1

177

Table 7.6   (continued)

Species	Lomond	Awe	Ness	Morar	Shiel
Oreodytes septentrionalis	1	0	0	0	0
Esolus parallelepipedus	0	0	6	0	0
Limnius volkmari	41	0	0	0	0
Agapetus fuscipes	8	0	4	0	0
Philopotamus montanus (Donovan)	0	0	0	2	0
Plectrocnemia geniculata McLachlan	1	0	0	0	0
Polycentropus flavomaculatus	3	1	12	0	0
Cyrnus flavidus	1	1	0	0	0
Tinodes waeneri	9	0	10	0	0
Phryganea varia	0	0	2	0	0
Lepidostoma hirtum	4	0	44	0	1
Sericostoma personatum	3	0	4	0	0
Total	1511	461	1188	5	11
Total number of species	36	20	28	3	5
Number of samples	4	5	3	2	2

Fauna-substrate associations

The littoral substrates of the shoreline of each loch fell into three main categories: silt and sand, gravel, and stones and boulders (Plate 7.2). Sites very rarely consisted of only one type of substrate and were assigned to each category according to the highest percentage of material that occurred there. The faunas of these different substrates are given in Tables 7.5, 7.6 and 7.7.

The most varied fauna occurred in Loch Lomond on stony substrates and the most abundant occurred in Awe on silt and sand. Lochs Lomond and Ness also had their most abundant faunas on silty/sandy substrates. All five lochs had their least abundant and (apart from Loch Ness) least varied faunas in gravel. Loch Ness was probably atypical because of enrichment due to sewage at one gravel site. It is interesting to note too that it was at this site (see below) that *Phagocata woodworthi* was collected in relatively large numbers.

There were similar patterns of occurrence in all five lochs when species inhabiting the different substrates were compared. Tricladida such as *Planaria torva* and *Dendrocoelum lacteum* were more abundant on stony shores as

178

Table 7.7 Littoral zoobenthos identified to species (mean number per 10 min collection) from the stony shores of Lochs Lomond, Awe, Ness, Morar and Shiel. Authorities are inclued only for those species not in Table 7.4

Species	Lomond	Awe	Ness	Morar	Shiel
Planaria torva	8	0	0	0	0
Dendrocoelum lacteum	3	0	0	0	0
Potamopyrgus jenkinsi	5	0	0	0	0
Lymnaea truncatula	0	3	1	0	0
L. peregra	10	10	16	0	0
Physa fontinalis	208	0	0	0	0
Planorbis laevis	2	0	0	0	0
P. albus	4	0	0	0	0
Ancylus fluviatilis	2	2	0	0	0
Stylaria lacustris	119	12	6	0	1
Lumbriculus variegatus	86	21	15	5	6
Stylodrilus heringianus	56	33	7	0	0
Eiseniella tetraedra	7	5	5	0	0
Glossiphonia complanata	4	0	1	0	0
Helobdella stagnalis	8	0	0	0	0
Erpobdella octoculata (L.)	0	1	0	0	0
Asellus aquaticus	121	0	0	0	0
Gammarus lacustris	0	46	0	0	0
G. pulex	9	35	0	0	0
Ameletus inopinatus	0	0	3	0	0
Centroptilum luteolum	21	4	0	0	0
Cloeon simile	1	1	0	0	0
Ecdyonurus venosus	31	0	49	0	0
E. torrentis Kimmins	0	1	0	0	0
E. dispar	1	1	0	9	6
Leptophlebia marginata	4	1	0	0	0
L. vespertina	3	0	0	0	0
Ephemerella ignita	0	0	0	1	0
Ephemera danica	4	1	0	0	0
Caenis moesta	768	0	0	0	0
C. horaria	115	0	0	0	0
Amphinemura sulcicollis	1	0	57	0	0
Nemoura avicularis	23	0	0	0	0
N. cinerea	0	4	0	0	0
Leuctra inermis	0	0	0	0	1
L. hippopus	0	21	38	0	4
L. nigra	0	1	0	0	0
L. fusca	0	0	0	0	9
Capnia bifrons	0	10	36	0	0
C. atra	7	0	0	0	0
Diura bicaudata	2	15	42	1	0
Chloroperla torrentium	99	7	100	0	0

179

Table 7.7 (continued)

Species	Lomond	Awe	Ness	Morar	Shiel
C. tripunctata	4	0	18	2	10
Ischnura elegans	2	0	0	0	0
Micronecta poweri	0	2	0	0	0
Sigara distincta	6	0	0	0	0
Deronectes depressus	4	3	0	0	0
Esolus parallelepipedus	1	0	4	0	0
Limnius volkmari	5	0	4	3	1
Agapetus fuscipes	3	0	1	0	0
Plectrocnemia conspersa	0	1	0	0	0
Polycentropus flavomaculatus	4	1	2	0	0
Tinodes waeneri	4	0	114	0	0
Lepidostoma hirtum	9	0	24	0	1
Sericostoma personatum	1	0	1	0	0
Total	1775	242	544	21	39
Total number of species	41	26	22	6	9
Number of samples	9	11	9	18	19

were some Gastropoda e.g. *Physa fontinalis* and *Ancylus fluviatilis*. In contrast, burrowing Oligochaeta were less abundant at stony sites.

Of the Crustacea, Cladocera were almost entirely restricted to the sheltered waters of the silty/sandy shores. Thus large numbers of *Eurycercus lamellatus* occurred at Loch Awe on silty/sandy shores but were absent from shores of stones and gravels. *Asellus* spp. too were more numerous at these sheltered sites while in contrast *Gammarus* spp. (Plate 7.5a) were far more abundant on stony and gravelly shores.

In a similar way, the species of mayflies inhabiting different substrates can be divided into two groups. In the silty/sandy substrates *Siphlonurus lacustris*, *Centroptilum luteolum*, *Cloeon* spp. and *Leptophlebia* spp. were prevalent. The stony substrates were characterized by Ecdyonuridae and *L. marginata*. *Ephemera danica*, *Caenis moesta* and *C. horaria* (Plate 7.5b) were found on all substrates including stony shores (presumably where silt and sand were occurring between stones).

Stoneflies were more abundant and varied in the stony substrates. There was an almost total lack of Leuctridae (Plate 7.5c) on the silty/sandy shores and very few *Chloroperla* sp. or *Diura bicaudata;* the main species on this type of substrate was *Nemoura avicularis.*

The distribution of water-bugs was almost the reverse of that of stoneflies. Only *Micronecta poweri* (Plate 7.5d) was found on stony shores and all the other species were very largely restricted to the sheltered silty areas. *Sigara dorsalis* was particularly common and abundant at these sheltered sites.

Megaloptera (in the form of *Sialis lutaria* in Lochs Lomond and Ness) and Odonata *(Ischnura elegans* in Loch Lomond) were not common and tended to be restricted to sheltered sites. Too few *S. lutaria* were collected from Loch Lomond for them to appear in the Tables.

Of the various species of beetles collected, elminthids were common and abundant on stony shores. Although *Deronectes depressus* was collected from silt at Loch Ness it was more usually found in gravel and stones (e.g. at Lochs Lomond and Awe). *Haliplus variegatus, Orectochilus villosus* and *Coleambus novemlineatus* occurred exclusively at the silty/sandy sites and *H. fulvus* too was more abundant there.

Most of the identified caddis species seemed to prefer the stony shores, except for *Cyrnus flavidus* which was more common on the sandy/silty and gravel shores. Phryganeidae did not occur on stony shores.

Seasonality

The similarity of the substrate of the seasonal sampling sites can be seen in Table 7.9 and Plate 7.4. It mainly consisted of stones ranging from 60% at

Table 7.8 Littoral zoobenthos identified to species (weighted mean number per 10 min sample) collected during the seasonal sampling of 1977–1978 from Lochs Lomond, Awe, Ness, Morar and Shiel. Authorities are included only for those species not in Table 7.4

Species	Lomond	Awe	Ness	Morar	Shiel
Lymnaea truncatula	0	3	0	0	0
L. peregra	6	27	96	2*	15*
Ancylus fluviatilis	4	3	1*	0	3*
Stylaria lacustris	19	7	73	9*	88
Lumbriculus variegatus	1	10	8	1	13
Stylodrilus heringianus	2	29	15	2*	17*
Eiseniella tetraedra	1	2	3*	0	0
Glossiphonia complanata	0	1*	5	0	0
Eurycercus lamellatus	0	0	0	0	1*
Polyphemus pediculus L.	0	9*	0	0	0
Gammarus lacustris	0	121	0	0	0
G. pulex	3	28	0	0	0

181

Table 7.8 (continued)

Species	Lomond	Awe	Ness	Morar	Shiel
Siphlonurus lacustris	0	0	0	0	5*
Ameletus inopinatus	0	0	23	13*	0
Baetis rhodani	0	0	0	3*	0
B. muticus (L.)	1*	0	0	0	0
Centroptilum luteolum	4	9	37*	4*	1*
Heptagenia lateralis (Curtis)	0	1*	4*	1*	0
Ecdyonurus venosus	40	0	88	11*	0
E. torrentis	0	37*	0	0	0
E. dispar	23	158	70*	20	0
Leptophlebia marginata	0	0	0	0	13*
Leptophlebia vespertina	0	7*	0	0	0
Ephemerella ignita	68*	288*	0	19	0
Ephemera danica	0	1	0	0	0
Amphinemura sulcicollis	1	0	132	3*	1*
Nemoura avicularis	0	0	0	0	1*
N. erratica Classen	1*	0	0	0	0
Leuctra inermis	1*	0	59	1*	0
L. hippopus	0	0	29	2*	1
L. nigra	1*	3	0	0	0
L. fusca	4*	81*	130*	12*	79
Capnia atra	0	0	0	2*	13*
C. bifrons	0	21	3	0	0
Diura bicaudata	21	35	46	4	0
Chloroperla torrentium	13	37	245	16*	79
C. tripunctata	0	0	14	4	0
Micronecta poweri	0	1	0	0	0
Deronectes depressus	2	5	0	0	3*
Oreodytes rivalis (Gyllenhal)	0	0	2*	1*	0
O. septentrionalis	5	1*	0	0	1*
Esolus parallelepipedus	0	4*	29	0	0
Limnius volkmari	3	1*	26	2	0
Agapetus fuscipes	0	0	4	0	0
Polycentropus flavomaculatus	1	20	12 .	2*	1*
Tinodes waeneri	11	2*	15	2*	22*
Lepidostoma hirtum	8	4*	11	1*	0
Sericostoma personatum	1	1*	2	0	3*
Total	245	957	1182	137	360
Total number of species	26	32	28	24	20

* Species which are additional to those presented in Table 7.4.

Plate 7.4 Predominant substrates (bar length represents 10 cm) at the seasonal sampling sites (a) Lomond (b) Awe (c) Ness (d) Morar (e) Shiel. (Photos: B. D. Smith, P. S. Maitland.)

Loch Shiel to 85% at Loch Ness. Gravel usually occurred between the stones, and boulders dotted the shore at all sites. Areas of sand occurred at Lochs Lomond, Awe and Shiel, and there were also small areas of silt at the latter site.

The fauna inhabiting these stony shores was generally similar at all lochs, as can be seen in Table 7.8. If the sites are ranked in decreasing order of species diversity and abundance then Loch Awe and Loch Ness are respectively first followed by Lochs Shiel, Lomond and Morar. It is clear that the north basin site of Loch Lomond was considerably impoverished compared with the south basin sites. This Lomond site was also probably more exposed (the stones being rounded) than those at other lochs.

The seasonal abundance of six zoobenthos species that were common in all

Table 7.9 Comparison of substrates at the seasonal sampling sites (site number in parenthesis)

Substrate		Lomond (45)	Awe (51)	Ness (3)	Morar (32)	Shiel (34)
Silt (0.063 mm)	%	0	0	0	0	5
Sand (0.063 mm–2.00 mm)	%	5	2	0	0	15
Gravel (2.00 mm–6.40 cm)	%	5	10	2	5	10
Stones (6.40 cm–25.60 cm)	%	80	78	85	70	60
Boulders (> 25.60 cm)	%	10	10	13	25	10

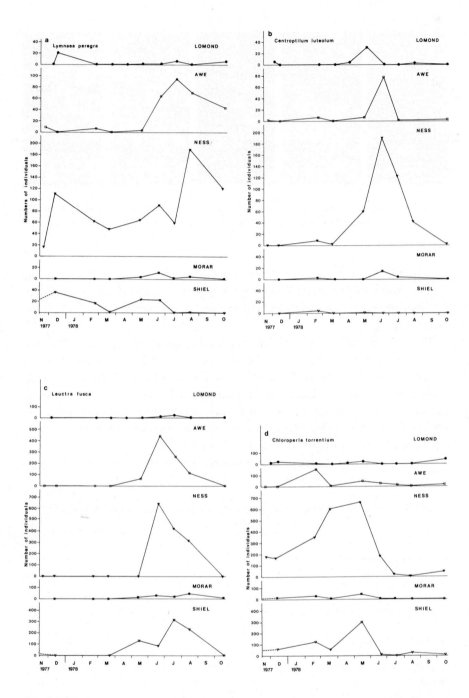

Fig. 7.2a-d

184

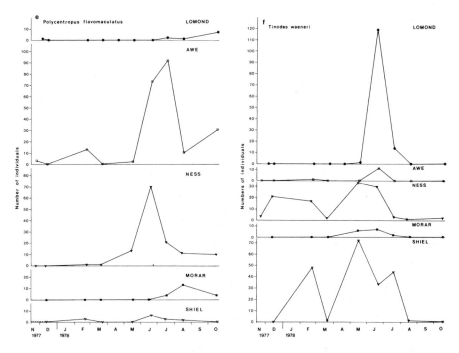

Fig. 7.2 The seasonal abundance of – (a) *Lymnaea peregra*, (b) *Centroptilum luteolum*, (c) *Leuctra fusca*, (d) *Chloroperla torrentium*, (e) *Polycentropus flavomaculatus* and (f) *Tinodes waeneri*, in the large lochs, numbers of individuals per 10 min collection. (The dashed line, where shown, indicates the value for Loch Shiel on the 23.9.1977 and Loch Morar on the 2.10.1977).

five lochs is given in Fig. 7.2a-f, and all are characteristic of the stony shores of oligotrophic lakes.

Lymnaea peregra was more abundant in Loch Ness than in any other loch (mean number: 96/10 min collection). Lochs Awe and Shiel had considerably fewer (27 and 15/10 min respectively) and very few were collected from Loch Morar (2/10 min) or the north basin site at Loch Lomond (6/10 min). High water levels in some months (e.g. at Lochs Morar and Shiel during October 1978) accounted for the absence of snails at these times. Large numbers of eggs were observed in Loch Ness in May and in Loch Awe in June. The increase in numbers of snails in Loch Awe (May-June) and Loch Ness (July-August) was due to the large numbers of spats present (see Fig. 7.3b). Numbers declined at most sites after August.

Centroptilum luteolum was most abundant at Loch Ness (mean number 37/10 min collection) followed by Lochs Awe (9/10 min collection), Lomond and Morar both 4/10 min). Very few individuals were collected in Loch Shiel

(1/10 min). Numbers started increasing in Loch Lomond in May, whereas in Lochs Ness, Morar and Awe increases did not occur until June. Numbers rapidly declined after these times.

Leuctra fusca nymphs occur mostly during the summer months. It was collected in Lochs Morar and Shiel during late September/early October 1977, as the 'dashed' line on Fig. 7.2c indicates, but was not collected in December 1977. It was particularly abundant in Lochs Shiel (mean number: 79/10 min collection), Awe (81/10 min) and Ness (130/10 min). It first started increasing in numbers in Loch Shiel, between March and May. (A few specimens were also collected from Loch Awe in May). Large increases in numbers occurred in Lochs Awe and Ness from May to June and numbers declined rapidly after this. Few *Leuctra fusca* were collected in Loch Morar (mean number 12/10 min) and even fewer in Loch Lomond (4/10 min). Increases in numbers did not occur in Lochs Lomond and Morar until July-August. By mid-October only a few specimens were collected and these all occurred in Loch Awe.

Chloroperla torrentium was particularly abundant in Loch Ness (mean number: 245/10 min collection), and moderately abundant in Lochs Shiel (79/10 min) and Awe (37/10 min). It was considerably less numerous at Lochs Morar (16/10 min) and Lomond (13/10 min). Peak numbers of *C. torrentium* tended to occur earlier than the other species discussed, with an increase in numbers from December to February in Lochs Awe, Ness and Shiel. Numbers decreased rapidly in Loch Awe after February and after May in Lochs Ness and Shiel. Numbers increased again in Loch Shiel in August but in the remaining four lochs only a single specimen was collected (at Loch Ness) during August. Small peaks occurred in Lochs Morar and Lomond during May, thereafter numbers decreased as at the other lochs.

Polycentropus flavomaculatus was collected most frequently in Loch Awe (mean number: 20/10 min collection) and Ness (12/10 min collection), with large increases in numbers occurring between May and June at both lochs. Numbers decreased rapidly after these maxima had been reached. Small peaks occurred in Loch Shiel in June, in Loch Lomond in July and in Loch Morar during August, thereafter numbers decreased.

Tinodes waeneri, in contrast to the preceding species of caddis, was collected in relatively high numbers from Lochs Lomond (mean number: 11/10 min collection) and Ness (15/10 min). However the highest numbers were collected in Loch Shiel (22/10 min). In Lochs Ness and Shiel the numbers were high for most of the year (the apparent decreases in March were probably due to the high water levels of that month). Peaks occurred during May and numbers decreased thereafter, very few animals being collected

186

during August and October. A very large maximum was reached in Loch Lomond in June but numbers declined rapidly after this. Few *T. waeneri* occurred in Lochs Awe and Morar, both had a mean number of 2/10 min.

Life cycles and growth

All the species had simple univoltine life cycles with adults dying soon after oviposition. *Lymnaea peregra* and *Chloroperla torrentium* occurred in all five lochs and *Diura bicaudata* in all except Loch Shiel. The mayflies *Ephemerella ignita* occurred in Lochs Lomond, Awe and Morar and *Ameletus inopinatus* in Lochs Ness and Morar. Mean body length values of each sample are plotted for the five species in Fig. 7.3a-7.7a and the range of size within the population is shown in bar charts in Fig. 7.3b-7.7b.

Lymnaea peregra. The growth of *Lymnaea peregra* is shown in Fig. 7.3a and 7.3b. The specimens from Loch Awe grew faster and attained a greater height (19 mm) than in any other of the lochs. The next largest were from Lochs Lomond (13 mm) and Ness (11 mm). Specimens from Lochs Morar and Shiel were always small and thin shelled, the largest measuring only 9 and 6 mm respectively. Young snails (*ca.* 2 mm) occurred first in Loch Lomond in July, in Loch Awe in August and in Loch Ness in October. (Samples were not collected during September 1978). The growth of snails in Loch Shiel was consistently slower than in any other loch with no growth apparent during the winter. There was similarly little growth during the winter at Loch Ness. Increases in the growth rate occurred in all lochs during the spring and early summer.

Ameletus inopinatus was fairly abundant in Lochs Ness and Morar but more numerous in Loch Ness (mean number: 23/10 min collection) than in Loch Morar (13/10 min). Maximum numbers were collected from both lochs during May and numbers declined rapidly after this month. Individuals at both lochs reached maximum lengths of 11 mm and growth rates appeared to be very similar, particularly during the winter and early spring when continuous growth occurred in both lochs. However, growth seemed to be faster in Loch Ness during the late spring and early summer with specimens of 11 mm in length occurring in May. Once emergence started the mean size of the population fell owing to the loss of the larger mature nymphs. Thus there was a drop in mean size from May to June. *Ameletus* 11 mm long were not collected from Loch Morar until August when a single specimen was found. Small specimens were collected in Loch Ness in fairly high numbers (41/10 min collection) during October. No small specimens were collected from

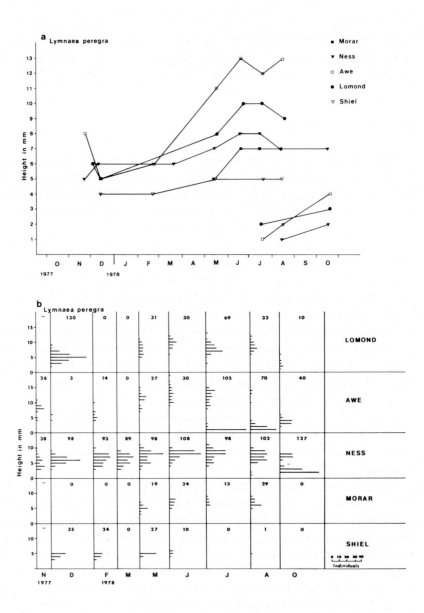

Fig. 7.3

a. The mean body length to the nearest mm, of each sample of *Lymnaea peregra* collected from Lochs Lomond, Awe, Ness, Morar and Shiel.

b. Size-structure bar charts of *Lymnaea peregra* in Lochs Lomond, Awe, Ness, Morar and Shiel. Each sample is divided into 1 mm size categories, the width of each size group representing the number of individuals found in that particular group. The numbers in each sample are indicated.

188

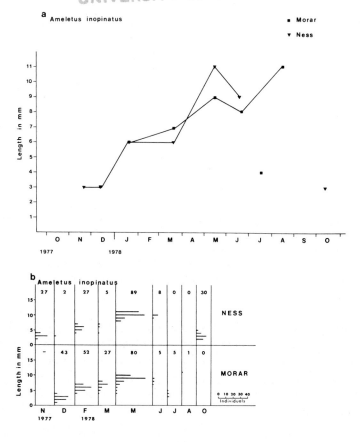

Fig. 7.4

a. The mean body length to the nearest mm, of each sample of *Ameletus inopinatus* collected from Lochs Ness and Morar.

b. Size-structure bar charts of *Ameletus inopinatus* in Lochs Ness and Morar. Each sample is divided into 1 mm size categories, the width of each size group representing the number of individuals found in that particular group. The numbers in each sample are indicated.

Loch Morar at this time. However, due to the particularly high water level of Loch Morar during October 1978, the total sample collected was small.

Ephemerella ignita. Nymphs of this species are generally absent during the winter months, although odd specimens were collected in Loch Lomond as late as December 6, 1977. However, only the nymphs collected during 1978 have been included in this comparison (Fig. 7.5a and 7.5b). This species first appeared in Lochs Awe and Morar in mid-May, but was not collected from Loch Lomond at that time. This is surprising, as the growth of *E. ignita* is

189

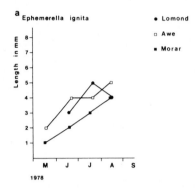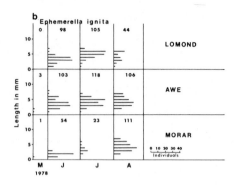

Fig. 7.5

a. The mean body length to the nearest mm, of each sample of *Ephemerella ignita* collected from Lochs Lomond, Awe and Morar.

b. Size-structure bar charts of *Ephemerella ignita* in Lochs Lomond, Awe and Morar. Each sample is divided into 1 mm size categories, the width of each size group representing the number of individuals found in that particular group. The numbers in each sample are indicated.

similar in Lochs Awe and Lomond though nymphs emerged earlier from Loch Awe. The largest specimen collected (8 mm) was also from Loch Awe.

The characteristic drop in mean population size had not occurred in Loch Morar even by mid-August. Samples were not collected during September and no specimens of *E. ignita* were collected in October. Presumably emergence took place during the late summer – early autumn in this loch. Samples taken during the previous autumn from Loch Morar (October 2, 1977) contained specimens, up to 10/10 min collection. The growth of this species was consistently slower in Loch Morar than in Lochs Lomond and Awe.

This species was seasonally very abundant, occurring in numbers of up to 681/10 min collection in Loch Lomond (in June) and 2240/10 min in Loch Awe (in June). Numbers were high in these two lochs in July also, but declined rapidly after that. In Loch Morar the maximum number was reached in August and at 128/10 min was comparatively high for that loch.

Diura bicaudata. The results of growth studies of *Diura bicaudata* are given in Fig. 7.6a and 7.6b. *D. bicaudata* appears to grow consistently fastest in Loch Awe followed by Lochs Lomond and Ness and finally Loch Morar but the rates in all the lochs were rather similar. Largest specimens collected were 20 mm (Lomond), 17 mm (Awe), 16 mm (Morar) and 15 mm (Ness). Emergence seemed to take place during the spring months (March-May) starting earlier than the two previous mayfly species. Small nymphs were collected from all lochs in June and in Loch Awe were already up to 6 mm

190

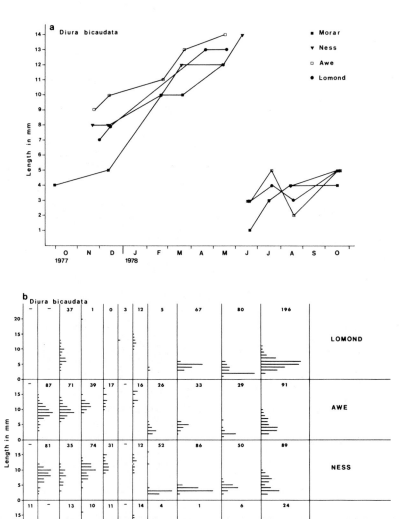

Fig. 7.6
a. The mean body length to the nearest mm, of each sample of *Diura bicaudata* collected from Lochs Lomond, Awe, Ness and Morar.
b. Size-structure bar charts of *Diura bicaudata* in Loch Lomond, Awe, Ness and Morar. Each sample is divided into 1 mm size categories, the width of each size group representing the number of individuals found in that particular group. The numbers in each sample are indicated.

191

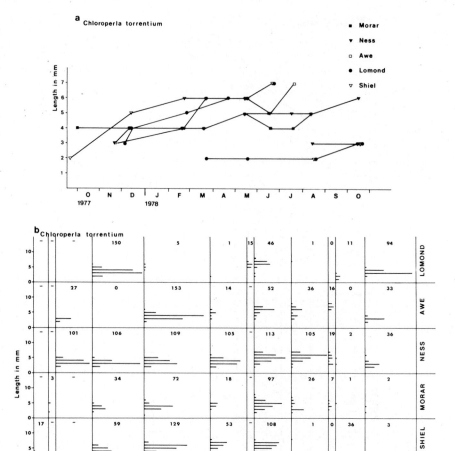

Fig. 7.7

a. The mean body length to the nearest mm, of each sample of *Chloroperla torrentium* collected from Lochs Lomond, Awe, Ness, Morar and Shiel.

b. Size-structure bar charts of *Chloroperla torrentium* in Lochs Lomond, Awe, Ness, Morar and Shiel. Each sample is divided into 1 mm size categories, the width of each size group representing the number of individuals found in that particular group. The numbers in each sample are indicated.

long. At the other end of the scale, in Loch Morar, the largest nymphs collected in the June sample were 2 mm long. By October there were nymphs of 10 mm or longer in all the lochs. Their growth rate did not appear to be slower during the winter.

Chloroperla torrentium. Fig. 7.7a and b show this species' growth curves and size distribution respectively. As with *D. bicaudata*, the growth rates (of *C. torrentium*) were similar in all five lochs. The fastest growth occurred in Loch Shiel, closely followed by that in Lochs Awe and Lomond. Lochs Ness and Morar had similar, slightly slower, growth rates. Very small nymphs were found as early as mid-March in Loch Lomond but it is uncertain if these represented the start of a new generation of *Chloroperla torrentium*. Very small nymphs are not found in the other lochs until August and October. Emergence seemed to be largely complete by August when very few *C. torrentium* were collected. In those lochs in which *C. tripunctata* also occurs, August is the only time of year that it is more numerous than *C. torrentium*.

Distribution related to depth

The depth distributions of certain species from each loch are shown in Table 7.10. Species were selected on the basis that they were reasonably abundant and occurred in three or more lochs. (The total numbers of animals used to calculate these percentages are also given in this table).

The animals fall into three main categories: those with preference for shallow water, those with a preference for deeper water and finally, the majority, those whose distribution does not appear to be related to water depth.

Belonging to the first group are *Ephemerella ignita*, *Ecdyonurus dispar* and possibly *Lepidostoma hirtum* (though in the last example the samples are rather small and the pattern not very marked). However the preference of the first two species for shallow water is striking. *Polycentropus flavomaculatus* and possibly *Deronectes depressus* have a preference for deeper water. Generally the remaining species, which include *Diura bicaudata* and *Chloroperla torrentium*, *Ecdyonurus venosus* and *Lymnaea peregra* do not appear to have preference for any particular depth. (Although *D. bicaudata* and *C. torrentium* were more abundant in the top 30 cm of water in Lochs Ness and Shiel).

E. venosus did not occur in Loch Awe but its place appeared to be taken by *E. torrentis*. Unlike the former species, *E. torrentis* did appear to have a preference for shallow water, much like *E. dispar*.

The occurrence of two species of *Gammarus* (Plate 7.5a) in Loch Awe, both in reasonably high numbers, is rather unusual and of considerable ecological interest. The depth distribution of *G. pulex* and *G. lacustris* in Loch Awe is given in Table 7.11 and compared with that of *G. pulex* in Loch Lomond. *G. lacustris* did not appear to be restricted or have a preference for

Table 7.10 The percentage distribution from December 1977–October 1978 at depths of 0–50 cm of various species from Lochs Lomond, Awe, Ness, Morar and Shiel (total numbers of animals are given in parentheses)

Loch	Depth in cm	Lymnaea peregra %	Ecdyonurus dispar %	Ecdyonurus venosus %	Ephemerella ignita %	Diura bicaudata %	Chloroperla torrentium %	Deronectes depressus %	Polycentropus flavomaculatus %	Lepidostoma hirtum %
Lomond	10	0	22	4	57	10	8	0	0	0
	20	25	43	23	24	25	16	0	0	0
	30	25	13	26	12	25	24	29	0	20
	40	0	15	23	4	23	24	29	20	40
	50	50	7	24	3	17	28	42	80	40
		(20)	(247)	(556)	(1566)	(334)	(169)	(33)	(19)	(26)
Awe	10	30	43	35*	71	25	10	0	4	0
	20	20	26	37*	19	35	29	6	22	38
	30	15	19	21*	6	23	14	41	9	12
	40	22	9	6*	3	7	16	29	26	12
	50	13	3	1*	1	10	31	24	39	38
		(233)	(858)	(319)	(2709)	(285)	(294)	(41)	(137)	(39)
Ness	10	31	29	30	100	35	39	100	5	19
	20	22	31	24	0	24	23	0	14	31
	30	17	24	18	0	20	23	0	9	19
	40	16	11	14	0	14	9	0	27	19
	50	14	5	14	0	7	6	0	45	12
		(885)	(402)	(493)	(2)	(380)	(2044)	(1)	(113)	(76)
Morar	10	12	36	44	39	17	28	0	10	0
	20	19	20	12	27	35	26	0	40	43
	30	25	20	20	15	17	18	100	30	14
	40	25	12	8	13	17	13	0	20	14
	50	19	12	16	6	14	15	0	20	29
		(49)	(377)	(100)	(461)	(165)	(264)	(1)	(23)	(48)
Shiel	10	0	—	—	—	—	38	33	25	40
	20	24	—	—	—	—	31	0	25	20
	30	29	—	—	—	—	18	0	0	20
	40	29	—	—	—	—	6	33	25	20
	50	18	—	—	—	—	7	34	25	0
		(106)					(596)	(19)	(12)	(28)

194

Table 7.11 The percentage distribution at depths of 0–50 cm of *Gammarus pulex* and *Gammarus lacustris* from Lochs Lomond and Awe (total numbers of animals are given in parentheses)

	Depth in cm	*Gammarus pulex* %	*Gammarus lacustris* %
Lomond	10	14	—
	20	37	—
	30	14	—
	40	9	—
	50	26	—
		(51)	
Awe	10	48	20
	20	23	25
	30	16	19
	40	8	18
	50	5	18
		(186)	(921)

any particular depth and this also appeared to be the case with *G. pulex* in Loch Lomond. However, *G. pulex* in Loch Awe was more abundant in shallow water.

Discussion

The most significant feature of the zoobenthos communities is the oligo-trophy of Lochs Morar, Shiel, Ness, Awe and the north basin of Lomond. When the overall data are compared then Loch Lomond clearly has the most abundant and varied fauna and Loch Morar the least. Lochs Awe and Ness occur somewhere between these extremes while Loch Shiel is next to Loch Morar in the general paucity of its fauna. These results agree with what is known about the chemistry, physical characteristics and substrates of these five lochs (see Tables 7.1 and 7.2).

If the relationships among common species (i.e. those occurring at more than one loch) are examined then each loch has more in common with Loch Lomond than any other loch – though Loch Morar has as many species in common with Loch Ness. This is not surprising considering the diversity of habitat in Loch Lomond and the extent of its species list. The similarity of Lochs Awe and Lomond has already been commented upon and these lochs have the highest number of common species. The similarities between Lochs

195

Ness and Lomond, and Ness and Awe have likewise been compared. Considering these physico-chemical factors it is not surprising that the fauna of Loch Morar is least like that of Loch Awe. What is surprising is that Lochs Shiel and Morar have so few species in common. This may indicate the importance of substrate with regard to the distribution of zoobenthos as the other physico-chemical factors are very similar in these lochs. Thus Loch Shiel has more species in common with Loch Lomond: the similarity of the substrates of these two lochs can be seen in Table 7.2.

The species of zoobenthos collected are typically those of oligotrophic waters and in many cases are generally considered to favour lotic habitats. The stony, wave-washed shores of such lochs aproximate to eroding rivers, (though truly lotic forms such as *Simulium*, *Rhithrogena* and *Hydropsyche* are absent apart from occasional specimens during spates). The lentic species occur only in sheltered bays where deposits of fine sediment accumulate. Calciphiles are largely absent.

Within these broad generalisations however, each loch is unique. Only fourteen of the species identified occurred in all five lochs. These were – *Lymnaea peregra*, *Stylaria lacustris*, *Lumbriculus variegatus*, *Stylodrilus heringianus*, *Centroptilum luteolum*, *Ecdyonurus dispar*, *Leuctra hippopus*, *Leuctra fusca*, *Chloroperla torrentium*, *Deronectes depressus*, *Limnius volkmari*, *Polycentropus flavomaculatus*, *Tinodes waeneri* and *Lepidostoma hirtum*. It is difficult to name a single species which is characteristic of each loch. However certain groupings and degrees of abundance do typify them.

Loch Lomond is characterized by the diversity and abundance of its fauna and the occurrence of species associated with productive waters. The few stoneflies (only 6% of the fauna of the shore-line survey) included *Nemoura avicularis*, numerous in productive lakes in the English lake district (Macan & Maudsley 1969). There were large numbers of *Asellus aquaticus*, a species often associated with organic-rich waters (Gledhill *et al.* 1976). *Gammarus pulex* was also present in much smaller numbers. Leeches (e.g. *Glossiphonia complanata*) and flatworms (e.g. *Dendrocoelum lacteum*), usually confined to productive lakes (Mann 1961, Reynoldson 1958), were common. Although all the snails collected from Loch Lomond were typical soft water species (Boycott 1936) there were more species there than in any other of the lochs.

At Loch Lomond too there was a wide variety of substrate types. The importance of substrates for zoobenthos distribution is well known (Moon 1934, 1936) and Loch Lomond has a full range of types from silty, reedy bays to exposed, wave-washed shores. Thus, to take mayflies as an example, the species list for Loch Lomond includes *Cloeon simile*, *Leptophlebia marginata*, *Caenis horaria* (habitat: silty reedy bays), *Emphemera danica* and

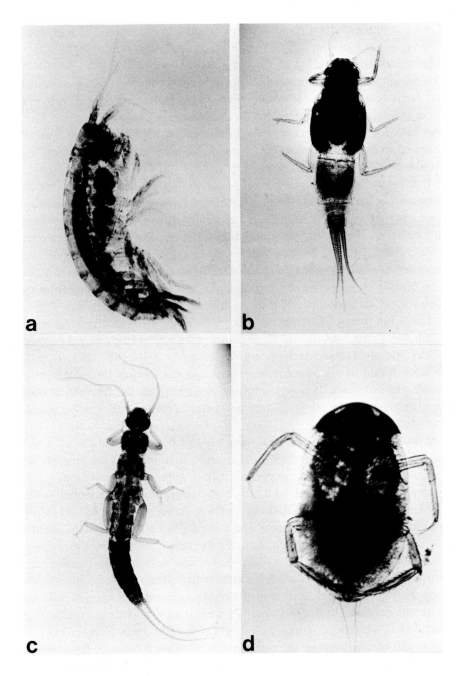

Plate 7.5 Examples of littoral zoobenthos inhabiting stony shores (a) *Gammarus* (b) *Caenis* (c) *Leuctra* (d) *Micronecta*. (Photos: P. J. Rickus.)

197

Caenis moesta (habitat: sand and gravel between stones) and *L. marginata, Centroptilum luteolum* and *Ecdyonurus* spp. (habitat: stony shores).

Much of the shoreline of Loch Lomond is wooded (71%). Weerekoon (1956a) suggested that the high productivity of the marginal zone was due to the large quantity of allochthonous vegetable matter entering it, providing food for detritivores such as *Asellus aquaticus, Caenis horaria* and *Ephemera danica*. In more recent years the nature and importance of such allochthonous material in freshwater systems has received much attention (e.g. Goldman 1961, Kaushik & Hynes 1968).

Loch Awe had the highest level of calcium of all the lochs and this was reflected in the relatively large numbers of *Lymnaea* spp. and *Gammarus* spp. and the abundance of the fauna generally. *Lymnaea peregra* grew faster in Loch Awe than in the remaining four lochs and the largest specimen was also collected there. There are trees along the shore which provide large amounts of allochthonous material. The fauna is less diverse than at Lochs Lomond or Ness. However, more unique species were collected from this loch than from any other and these included a wide range of leeches.

Large numbers of *Gammarus pulex* and *G. lacustris* occurred together at Loch Awe. This is unusual as *G. pulex* normally is found only near the mouths of streams when *G. lacustris* occurs in the same loch (Ratcliffe 1977). *G. pulex* is often found in oligotrophic reservoirs (*op. cit.*) where fluctuating water levels are common. It is possible that in Loch Awe the fluctuating water levels caused by the pumped-storage scheme favour *G. pulex* rather than *G. lacustris*.

Lymnaea truncatula occurred regularly at Loch Awe. This is not a truly aquatic species (Boycott 1936) and drowns if submerged for long periods (Hunter 1952). Its presence may again reflect water level fluctuations.

Loch Ness is characterized by the high densities and variety of stoneflies (30% of total shore-line fauna) there. These include *Capnia bifrons*, which was abundant during the winter and a number of species usually found in streams – *Nemoura cambrica, Leuctra hippopus* and *Chloroperla tripunctata*. The shores of Loch Ness slope steeply and wave action is considerable. Presumably this provides a habitat similar to that of small streams. A common mayfly in Loch Ness was *Ameletus inopinatus* a species which is generally confined to streams over 300 m but is known to occur in lochs in the far north of Scotland down to sea level (Macan 1979). Its occurrence in Lochs Ness and Morar must be the most southerly of such records. Its growth in both lochs was similar and like that found by Gledhill (1959) in a small stream in the English Lake District. However, specimens collected from Lochs Ness and Morar were never as large as those found by Gledhill.

Asellus aquaticus and *A. meridianus* occurred together at Loch Ness, this appears to be the first record of these species occurring together in Scotland. The presence in Loch Ness of large numbers of *Phagocata woodworthi*, – a species of North American triclad new to Britain – was of considerable interest. It seems likely that this species was introduced into the loch by American teams looking for the 'Loch Ness Monster' (Reynoldson *et al.* 1981). In Loch Ness *P. woodworthi* occurred mainly at a site enriched with sewage effluent and with large numbers of *Asellus* spp. and oligochaete worms. Prof. Reynoldson, who collected large numbers of live specimens for identification, was only able to find these in the vicinity of the sewage outflow (personal communication).

The most characteristic feature of both Lochs Morar and Shiel was the general sparsity of the fauna both in its abundance and diversity. However each loch is unique and surprisingly they have relatively few species in common. The substrate of the two lochs was different with large areas of sand and gravel in Loch Shiel and virtually none in Morar. Loch Morar had few trees on its banks while Shiel had about 37% cover. Recently, large areas of the catchment of Loch Shiel have been afforested with coniferous trees. Loch Morar was characterised more by the absence than presence of species. There were virtually no leeches, triclads and Hemiptera and only one species of mollusc – *Lymnaea peregra*. The growth of several species was slowest in Loch Morar and the individuals were usually smaller. Mayflies and stoneflies were among the most common animals. All the species are typical of stony, wave-washed shores or stony streams.

Ephemeroptera collected from Loch Shiel included *Siphlonurus lacustris* and *Leptophlebia marginata,* which generally occur in quieter water among weeds (Macan 1957). Very few Ecdyonuridae or *Centroptilum luteolum* occurred here and a notable stonefly absentee was *Diura bicaudata* (a species which was common and abundant in the other four lochs). *Chloroperla torrentium,* which occupies a similar niche, grows consistently faster in Loch Shiel than in any of the other lochs.

Fig. 7.8 is an attempt to summarise the invertebrate communities of these five large lochs and their relationship with one another. The main conclusions from this analysis are that, although the five lochs have many species in common, each loch has a characteristic community. Lochs Lomond, Awe and Ness are similar in that they each have large numbers of species, many of them in common. Lochs Morar and Shiel, on the other hand have many fewer species and have more in common with the other three lochs than with each other. They have few unique species, whilst Lochs Lomond, Awe and Ness each have six or more.

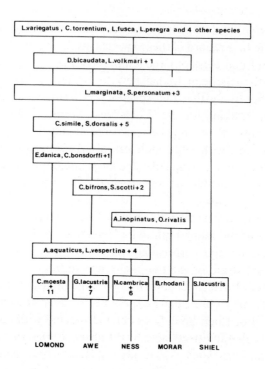

The boxes in the figure contain:

L.variegatus, C.torrentium, L.fusca, L.peregra and 4 other species

D.bicaudata, L.volkmari + 1

L.marginata, S.personatum +3

C.simile, S.dorsalis + 5

E.danica, C.bonsdorffi +1

C.bifrons, S.scotti +2

A.inopinatus, O.rivalis

A.aquaticus, L.vespertina + 4

| C.moesta + 11 | G.lacustris + 7 | N.cambrica + 6 | B.rhodani | S.lacustris |

LOMOND   AWE   NESS   MORAR   SHIEL

Fig. 7.8 The characterisation of the large lochs by selected zoobenthos species. This diagram was prepared from the data presented in Tables 4 and 8. It should be regarded as a general indication of the nature of the communities concerned and does not imply categorically that certain species *never* occur in some of the lochs from which they are not listed.

The pressures created by Man's activities on these lochs and their catchments have been discussed by Maitland (1981). Loch Lomond has more demands on its resources than any of the other lochs and it is fortunate that there have been other studies of the zoobenthos there, with which comparisons can be made. The similarity of the fauna described here with that found by Weerekoon (1956a) and Slack (1957) is marked. There is no evidence to suggest that Loch Lomond has changed markedly or deteriorated over the past thirty years.

As the present study has demonstrated, those lochs under the greatest pressure from Man (Lochs Lomond, Awe and Ness) have a far greater diversity and abundance of fauna compared with those least affected (Lochs Morar and Shiel). Further the growth rates and life cycles of species inhabiting Loch Awe and Ness appear to be unaffected by the operation of pumped-storage schemes on their shores. *Lymnaea peregra, Ephemerella ignita* and *Diura bicaudata* grow faster in Loch Awe than in any of the

200

remaining lochs in which they occurred. *Ameletus inopinatus* grew slightly faster in Loch Ness than in Loch Morar and reached its maximum size two months earlier. Invertebrates with a preference for shallow water would be the most adversely affected by serious water level fluctuations. The two species during the present study with the most distinct preference for shallow water, *Ecdyonurus dispar* and *Ephemerella ignita,* were more abundant at Lochs Awe and Ness (*E. dispar* only) than at the three remaining lochs without such schemes. These schemes appear to have negligible overall effects on the littoral zoobenthos of Lochs Awe and Ness, and the general conclusion is that, at the moment, the communities are controlled mainly by the more or less natural physico-chemical conditions found at each loch.

Acknowledgements

This study is part of a research programme carried out under contract to the Nature Conservancy Council and the North of Scotland Hydro-Electric Board. We are grateful to the following people who cheerfully assisted with field-work, during often difficult and arduous conditions – Mr K. H. Morris (who also helped with the identification of zoobenthos), Mrs P. Duncan, Mr A. J. Rosie, Mr A. A. Lyle, Ms G. M. Dennis and Mrs S. M. Adair. Further thanks are due to Mrs Adair for drawing the figures and to Mrs M. S. Wilson for typing the manuscript. Professor T. B. Reynoldson very helpfully identified all the Tricladida collected during this study. Finally we should like to thank Dr A. E. Bailey-Watts and Mr G. S. Scott for their useful comments on this paper.

References

Bailey-Watts, A. E. & Duncan, P., 1981. The ecology of Scotland's largest lochs: Lomond, Awe, Ness, Morar and Shiel. Ed. P. S. Maitland. Chemical characterisation – a one year comparative study. Chap. 3. This volume.

Boycott, A. E., 1936. The habitats of freshwater molluscs in Britain. J. Anim. Ecol. 5: 116–186.

Calow, P., 1974. Ecological notes on mayfly larvae (Ephemeroptera) in Loch Lomond. Glasg. Nat. 19: 123–130.

Chitty, A. J., 1892. Coleoptera at Loch Awe, June 1982, Ent. mon. Mag. 3: 216.

Chitty, A. J., 1893. Coleoptera at Loch Awe, June 1892 Ent. mon. Mag. 29: 48.

Fergusson, A., 1901. Order Coleoptera. Brit. Ass. Handb. Glasg. 272–301.

Fergusson, A., 1910. Additions to the list of Clyde Coleoptera (First Paper). Glasg. Nat. 2: 83–92.

Gledhill, T., 1959. The life history of *Ameletus inopinatus* Eaton. Hydrobiologia, 14: 85–90.

Gledhill, T., Sutcliffe, D. W. & Williams, W. D., 1976. A key to British Freshwater Crustacea: Malacostraca. Freshwater Biological Association Scientific Publication, No. 32.

Goldman, C. R., 1961. The contribution of alder trees *(Alnus terrui folia)* to the primary productivity of Castle Lake, California. Ecology, 42: 282–287.

Hunter, W. R., 1952. The adaptations of freshwater Gastropoda. Glasg. Nat. 16: 84–85.

Hunter, W. R., 1953a. On migrations of Lymnaea peregra (Müller) on the shores of Loch Lomond. Proc. R. Soc. Edinb. 65: 84–105.

Hunter, W. R., 1953b. The condition of the mantle cavity in two pulmonate snails living in Loch Lomond. Proc. R. Soc. Edinb. 65: 143–165.

Hunter, W. R., 1953c. On growth of the fresh-water limpet *Ancylus fluviatilis* Müller. Proc. zool. Soc. Lond. 123: 623–636.

Hunter, W. R., 1954 *Valvata piscinalis* (Müller) used by *Phryganea* in Loch Lomond. J. Conch. 23: 402–403.

Hunter, W. R., 1957. Studies on freshwater snails at Loch Lomond. In: Studies on Loch Lomond, 1. Ed. H. D. Slack. Glasgow: Blackie. 56–95.

Hunter, W. R., 1961a. Life cycles of four freshwater snails in limited populations in Loch Lomond, with a discussion of infra specific variation. Proc. zool. Soc. Lond. 137: 135–171.

Hunter, W. R., 1961b. Annual variations in growth and density in natural populations of freshwater snails in the West of Scotland. Proc. zool. Soc. Lond. 136: 219–253.

Hunter, W. R., Maitland, P. S. & Yeoh, P. K. H., 1964. *Potamopyrgus jenkinsi* in the Loch Lomond area, and an authentic case of passive dispersal. Proc. malac. Soc. Lond. 36: 27–32.

Hunter, W. R. & Slack, H. D., 1958. *Pisidium conventus* Clessin in Loch Lomond. J. Conch. 24: 245–247.

Hynes, H. B. N., 1961. The invertebrate fauna of a Welsh mountain stream. Arch. Hydrobiol. 57: 344–388.

Kaushik, N. K. & Hynes, H. B. N., 1968. Experimental study on the role of autumn-shed leaves in aquatic environments. J. Ecol. 56: 229–243.

Lawson, J. W. H., 1957. The biting midges and the dancing midges. In: Studies on Loch Lomond. 1. Ed. H. D. Slack. Glasgow: Blackie. 96–112.

Macan, T. T., 1957. The Ephemeroptera of a stony stream. J. Anim. Ecol. 26: 217–342.

Macan, T. T., 1979. A key to the nymphs of British species of Ephemeroptera with notes on their ecology. Freshwater Biological Association Scientific Publication, No. 20.

Macan, T. T. & Maudsley, R., 1969. Fauna of the stony substratum in lakes in the English Lake District. Verh. int. Verein theor. angew. Limnol. 17: 173–180.

Maitland, P. S., 1972. Loch Lomond: man's effect on the salmonid community. J. Fish. Res. Bd. Can. 29: 849–860.

Maitland, P. S., 1981. The ecology of Scotland's largest lochs: Lomond, Awe, Ness, Morar and Shiel. Ed. P. S. Maitland. Introduction and catchment analysis. Chap. 1. This volume.

Mann, H. P., 1961. The oxygen requirements of leeches considered in relation to their habitats. Verh. int. Verein. theor. angew. Limnol. 14: 1009–1013.

Moon, H. P., 1934. An investigation of the littoral region of Windermere. J. Anim. Ecol. 3: 8–28.

Moon, H. P., 1936. The shallow littoral region of a bay at the north west end of Windermere. Proc. zool. Soc. Lond. 2: 491–515.

Murray, J., 1910. Biology of the Scottish lochs. In: Bathymetrical survey of the freshwater lochs of Scotland. 1. Ed. J. Murray & L. Puller. Edinburgh: Challenger. 275–334.

North of Scotland Hydro-Electric Board, 1976. Report on Craigroyston pumped storage. Edinburgh: Unpublished report.

Ratcliffe, D. A., 1977. A nature conservation review. Cambridge University Press. Vols. 1 – 2.

Reynoldson, T. B., 1958. The quantitative ecology of lake dwelling triclads in Northern Britain. Oikos, 9: 94–138.

Reynoldson, T. B., Smith, B. D. & Maitland, P. S., 1981. A species of North American triclad new to Britain found in Loch Ness, Scotland. J. Zool. 193: 531–539.

Scott, T., 1899. The invertebrate fauna of the inland waters of Scotland – report on special investigation. Rep. Fishery Bd. Scot. 17: 133–204.

Scourfield, D. J., 1908. The biological work of the Scottish Lake Survey. Int. Revue ges. Hydrobiol. Hydrogr. 1: 177–192.

Sinclair, F. L., 1953. A preliminary list of stoneflies (Plecoptera) from the Glasgow area. Glasg. Nat. 17: 89–90.

Slack, H. D., 1957. The Fauna of the lake. In: Studies on Loch Lomond, 1, Ed. H. D. Slack. Glasgow: Blackie. 33–48.

Smith, I. R., Lyle, A. A. & Rosie, A. J., 1981. The ecology of Scotland's largest lochs: Lomond, Awe, Ness, Morar and Shiel. Ed. P. S. Maitland. Comparative physical limnology. Chap. 2. This volume.

Weerekoon, A. C. J., 1953. On the behaviour of certain Ceratopogonidae (Diptera). Proc. R. ent. Soc. Lond. A. 28: 85–92.

Weerekoon, A. C. J., 1956a. Studies on the biology of Loch Lomond. 1. The benthos of Auchentullich Bay. Ceylon J. Sci. 7: 1–94.

Weerekoon, A. C. J., 1956b. Studies on the biology of Loch Lomond. 2. The population of McDougall Bank. Ceylon J. Sci. 7: 95–133.

Wentworth, C. K., 1922. A scale of grade and class terms for clastic sediments. J. Geol. 30: 377–392.

8. The profundal zoobenthos

B. D. Smith, S. P. Cuttle & P. S. Maitland

Abstract

Little information was previously available on the sediments and invertebrate faunas of the deep waters of the large lochs of Scotland, except Loch Lomond. This paper reviews the published data and presents new information for the five largest lochs. The profundal sediments are characterised by very fine particle sizes (high fractions of clay and silt) and relatively high proportions of organic matter. The zoobenthos communities of the lochs are dominated by oligochaete worms (especially Tubificidae), sphaeriid molluscs (especially *Pisidium*) and various species of chironomid larvae. Only Loch Lomond has larvae of the culicid *Chaoborus*, thought to be an indicator of eutrophic conditions. Though many more comparative data are required from all the lochs, it is evident that the physico-chemical conditions and invertebrate communities in their abyssal regions are relatively similar to each other.

Introduction

The study of the profundal fauna of these large lochs was not one of the initial objectives of the project (Maitland 1981) and the principal field studies concentrated on water chemistry, phytoplankton (Bailey-Watts & Duncan 1981a, b), zooplankton (Maitland *et al.* 1981b) and the littoral fauna (Smith *et al.* 1981a). However, as with the macrophytes (Bailey-Watts & Duncan 1981c), it was decided to provide a brief account of the zoobenthos living in deep water in order to fill an obvious gap in the basic information on the ecology of the large lochs. This study is essentially a review one based on the literature, some unpublished data and a few recent samples obtained by the authors from the lochs concerned.

In terms of area, these large Scottish lochs are actually three or four orders of magnitude less than the really huge freshwater lakes of the world (Lakes Superior, Michigan, etc.). However, in terms of depth, the differences are much less and it is really the depth of the Scottish lochs that not only justifies describing them as large in a world context, but which also dominates many of their limnological and biological features. The profundal substrates and animals which live there permanently at depths exceeding 100, 200 or in one instance (Loch Morar) even 300 m are of considerable interest. The relatively large areas involved (about 50% of the area of Loch Ness lies in water deeper than 150 m) and the constancy of physical and chemical conditions at these depths make their habitats and communities rather distinct from those of the more superficial waters, to which most attention has been paid during the rest of this study

The earlier studies of the profundal sediments and fauna of these lochs were based on the extensive samples taken by Murray & Pullar (1910), especially in Loch Ness. Unfortunately they did not publish the results of their work in any detail, though they did present general lists of profundal zoobenthos species and some sediment analyses. They were initially unsuccessful in obtaining samples in Loch Morar and though they may have sampled in the other large lochs this is not made clear and no detailed results were ever published.

So far as is known, no details of the profundal sediments of Loch Awe have ever been published. On Loch Morar and Loch Shiel in 1968 Ekman grab samples were taken offshore (but not in really deep water) by N. C. Morgan (personal communication) as part of the Nature Conservation Review (Ratcliffe 1977). The data were never published but are available to the authors. Similar unpublished data are available from cores taken in deep water in Loch Morar as part of a survey there in 1969 by St. Andrews University (C. Muir, personal communication). There has been a considerable amount of research on the deep water sediments of Loch Morar in recent years (A. Duncan, A. Shine, personal communication) though none of it has yet been published.

The only large Scottish loch whose profundal sediments and animal communities has been studied in detail is Loch Lomond. Here, Slack (1954) carried out extensive studies in the bottom deposits and of the profundal fauna (Slack 1965). Additional data on the profundal fauna of Loch Lomond are available from Murray & Pullar (1910), Fedoruk (1964) and regular undergraduate studies there from Glasgow University, supervised by H. D. Slack and one of the present authors (Maitland 1978). The palynology of sediments in Loch Lomond has also been examined by Dickson *et al.* (1978). The only other large Scottish loch from which extensive data on sediments and zoo-

Plate 8.1 Sampling profundal zoobenthos at Loch Awe with the tow dredge used for collecting surface sediments in deep water. (Photo: A. E. Bailey-Watts.)

benthos are available is Loch Leven (Maitland 1979), but because of the shallowness of this water (mean depth 3.9 m, maximum depth 25.5 m) none of it can be properly described as profundal.

Methods

All sediments were collected near the stations where samples of water, zooplankton and phytoplankton were taken during the main part of the study. Full details are given by Smith *et al.* (1981b). A single core from the profundal area of Loch Ness was obtained in October 1978, using a Jenkin corer similar to that described by Mortimer (1941). Surface dredge samples were collected from Loch Lomond and Loch Awe on 2 October 1980, using a simple limnological tow dredge of the type described by Welch (1948). A net (mesh aperture: 1.0 mm) 35 cm long was attached to the mouth and enclosed in a protective coarse-mesh net (aperture: 10 mm) (Plate 8.1). The dredge was operated by lowering it to the bottom, paying out an amount of line equivalent to about 3 times the depth and then towing gently for a few minutes. It was then hauled as quickly as possible to the surface. The same dredge was used in November 1980 at Loch Shiel where a sample was successfully obtained and at Loch Morar where, apparently because of the great depth and the soft nature of the sediments involved, most of the mud was washed out through the net of the dredge by the time it reached the surface. Thus though some invertebrates were collected here, no satisfactory samples of sediment were obtained.

After collection, each sample of sediment was placed in an open container and mixed thoroughly. A sub-sample of about 0.5 l (for subsequent physical and chemical analyses) was placed in a labelled jar and dried in the laboratory. The remaining sediment was washed carefully in a pond net (mesh aperture: 0.5 mm) and the washed residue preserved in 4% formaldehyde. These samples were later sorted in the laboratory and all animals identified and counted. The sample from Loch Ness was sufficient only for physical and chemical analysis and no zoobenthos were obtained there.

After drying, the samples for physico-chemical analysis were passed through a 2 mm mesh sieve, coarser aggregates being crushed with a wooden pestle where necessary. The Loch Awe sample contained several large artefacts of rusted iron (representing 14% of the total sample weight) and these were removed prior to sieving.

The sieved samples were subjected to the following analyses, carried out in duplicate.

Particle size analysis. Particle sizes were fractionated on the basis of the Wentworth Scale; very coarse sand (2000–1000 μm), coarse sand (1000–500 μm), medium sand (500–250 μm), fine sand (250–125 μm), very fine sand (125–62.5 μm), silt (62.5–39 μm) and clay (< 39 μm) (Wentworth 1922).

The sieved sediment was treated with hydrogen peroxide to destroy organic matter and sodium hexametaphosphate added in the suspension as a dispersing agent. The clay and silt fractions were determined by the pipette method, employing settling times appropriate to the limiting particle diameters (Piper 1950). Not all the organic matter was destroyed by the initial hydrogen peroxide treatment and where organic material was present in the pipetted sample it was destroyed by further treatment with hydrogen peroxide prior to drying and weighing.

After removal of remaining silt and clay by decantation, the residue was dried and the coarser fractions determined by dry sieving on a mechanical shaker (Cummins 1962).

Moisture content and loss on ignition. The moisture content of the airdry sample was determined as the loss in weight after heating at 105 °C for 16 h. The sample was then ignited in a muffle furnace at 500 °C for 3 h and the weight loss determined.

pH. Sediment and water were mixed in a 1: 2.5 (w/v) ratio and stirred at intervals. After 1 h the pH of the freshly stirred suspension was measured using a pH meter and glass/reference electrode assembly.

Nitrogen content. The finely ground sample was digested with a concentrated sulphuric acid/hydrogen peroxide mixture containing a selenium catalyst. Ammonium-nitrogen was determined in the digest solution by an automated colorimetric procedure (Crooke & Simpson 1971).

Phosphorus, calcium, iron and silicon contents. After fusion of the finely ground sample with anhydrous sodium carbonate, the solidified melt was dissolved in dilute hydrochloric acid. The solution was evaporated to dryness and the residue heated overnight at 105 °C to dehydrate the silica. The residue was treated with dilute hydrochloric acid and the insoluble silica separated by filtration. Final traces of silica were removed from the filtrate by again evaporating to dryness, treating with dilute acid and filtering. Both filter papers containing the silica residue were ashed together and silica determined as the loss in weight of the ignited residue on treatment with hydrofluoric acid. Procedures for the sodium carbonate fusion and determination of silicon were adapted from Allen *et al.* (1974).

209

The filtrate from the silicon determination was evaporated to fuming with sulphuric acid, diluted and boiled for several minutes before cooling and adjusting to volume. Phosphorus was determined in this solution as ortho-phosphate using an automated colorimetric procedure, modified from Murphy & Riley (1962). Iron and calcium concentration in the solution were determined by atomic absorption spectroscopy, employing lanthanum as a releasing agent in the calcium determination.

Profundal sediments

Mean data from the physical and chemical analyses of the profundal sediment samples are presented in Table 8.1.

Results of the particle size analyses demonstrate that all the samples are composed mainly of the finer size fractions. Particles in the range of very coarse sand to fine sand constitute only a small part of the total and these fractions are therefore combined in the presentation of the data. Values are generally similar for all samples except that the sediment from Loch Ness contains a lower percentage of clay and correspondingly more material in the silt fraction and particularly in the very fine sand fraction.

The pH value determined for the Loch Awe sample is higher than those for the other lochs which are similar to one another. Samples from Lochs Awe

Table 8.1 Particle size and chemical analyses of profundal sediment samples from the large lochs. Particle size analysis: % by weight of mineral fraction, chemical analyses: % by weight of oven dry material

Loch	Lomond	Awe	Ness	Morar	Shiel
Very coarse – fine sand					
(125–2000 μm)	1.6	3.7	0.8	—	1.5
Very fine sand					
(63–125 μm)	15.4	12.6	35.4	—	17.4
Silt (39–63 μm)	9.0	6.9	18.1	—	15.3
Clay (< 39 μm)	74.1	76.7	45.8	—	65.7
pH	5.3	6.3	5.4	—	5.3
Loss on ignition	11.8	23.9	8.9	—	24.4
Nitrogen	0.33	0.78	0.24	—	0.75
Phosphorus	0.19	0.18	0.11	—	0.12
Calcium	0.24	0.96	0.94	—	1.15
Iron	9.5	10.5	4.1	—	4.1
Silicon	22	22	27	—	27

and Shiel give higher loss on ignition values and this is reflected in their higher contents of nitrogen. Phosphorus contents are highest in the samples from Lochs Lomond and Awe, the Lomond sample also having the highest content of iron but the lowest of calcium. The value for the concentration of iron in the Loch Awe sample should be treated with caution as the large amount of rusted material originally present in the sample makes interpretation difficult. Contents of silicon were slightly lower in the Loch Lomond and Awe samples than in the other lochs.

Profundal zoobenthos

The profundal fauna of all five lochs is basically composed of bivalve molluscs, oligochaete worms and chironomid larvae (Table 8.2). In some of the lochs other groups have been found, such as the culicine larva *Chaoborus*.

Loch Lomond

More information is available on the profundal fauna of Loch Lomond than on any of the four remaining lochs. An extensive survey was carried out by Slack (1965) during 1961 and 1962. Material was collected from depths ranging from 15 to 190 m. Four species of bivalve mollusc were identified (Table 8.2) *Pisidium lilljeborgi* and *P. nitidum* were found only in the south basin while *P. conventus* and *P. casertarnum* were found at all other sites. The oligochaetes collected were dominated by Tubificidae. *Tubifex* and *Peloscolex ferox* were rarely absent from any samples and it was these same two species which were found in the recent sample collected during October 1980 in the deep water (190 m) of the north basin. The third common oligochaete, *Limnodrilus*, found by Slack (1965) was not collected from depths of more than 60 m and was not found during the present survey.

The lumbriculids *Lumbriculus variegatus* and *Stylodrilus heringianus* were also collected by Slack (1965), the latter species was confined to deep water, and both species were collected during the present survey. Naid worms are uncommon in deep water apart from *Paranais* sp. and *Arcteonais lomondi* (Plate 8.2). The latter was originally found in the north basin and described by Martin (1907).

Chironomidae form an important part of the profundal fauna in Loch Lomond (Table 8.2). *Tanytarsus signatus, Tanytarsus* type A, *Procladius* spp., *Sergentia coracinus* and *Polypedilum convictus* were particularly

211

Table 8.2 The profundal macrozoobenthos of Lochs Lomond, Awe, Ness, Morar and Shiel (source of data indicated by initial)

				Lomond	Awe	Ness	Morar	Shiel
COELENTERATA			Hydra			B		
MOLLUSCA	Gastropoda							
			Lymnaea peregra (Muller)	S		B	M	L
	Pisidium			S		B		
			P. casertanum (Poli)	S		BO		
			P. conventus (Clessin)	S		O		
			P. personatum Malm			O		?L
			P. lilljeborgi Clessin	S				
			P. nitidum Jenyns	S				
OLIGOCHAETA	Naididae							
			Paranais	S				
			Arcteonais lomondi Martin	S			M	
	Tubificidae							
			Tubifex tubifex (Muller)	SL	L		M	ML
			Limnodrilus hoffmeisteri Claparede	?SL	L		M	L
			Peloscolex ferox (Eisen)	?S				?L
	Enchytraeidae			SL				L
	Lumbriculidae			SL		B		
			Lumbriculus variegatus (Muller)	SL			L	
			Stylodrilus heringianus Claparede	SL		B	L	
HIRUDINEA			Helobdella stagnalis (L)	S			L	
HYDRACARINA			Limnesia undulata (Muller)	S			M	
			Neumania callosa (Koenike)	S			M	
INSECTA	Ephemeroptera							
			Centroptilum luteolum (Muller)				M	
	Coleoptera							
			Dryops				M	

Family	Taxon	S	L	B	M	ML
Tipulidae	*Dicranota*				M	
Chaoboridae	*Chaoborus flavicans* (Meigen)	S			M	
Ceratopogonidae	*Probezzia*	S				
	Bezzia	S				ML
Chironomidae	*Macropelopia* sp	SL	L	B	M	
	Procladius	S	L			
	Ablabesmyia	SL				
	Protanypus morio (Zetterstedt)	S				
	Heterotanytarsus apicalis (Kieffer)	S				
	Orthocladius	S				
	Prodiamesia olivacea (Meigen)	S				
	Corynoneura	S				
	Metriocnemus	S	L			L
	Chironomus anthracinus (Zetterstedt)	S		B		
	Chironomus			B		
	Cryptochironomus	S	L			
	Endochironomus	S	L			
	Limnochironomus	S				
	Microtendipes	S				
	Pagastiella	S				
	Polypedilum	SL	L			L
	Sergentia	S				
	Stictochironomus	S				
	Micropsectra	S				
	Tanytarsus	S	L		M	
	Stempellinella	S				

S = Slack; L = Large lochs survey; B = Bathymetrical survey; M = Morgan; O = Oldham.

abundant. Twenty-eight other species were identified, but for the sake of simplicity in constructing this table, generally only the genus is indicated. A detailed list of the chironomids collected is given by Slack (1965). Other dipteran larvae collected by Slack were the culicid *Chaoborus flavicans* (Plate 8.3) and larvae of the ceratopogonid midges *Bezzia* and *Probezzia*.

Other minor elements of the fauna collected by Slack were unidentified enchytraeid worms (also found during the present survey), the leech *Helobdella stagnalis* and two Hydracarina *Limnesia undulata* and *Neumania callosa*.

Loch Awe

Information on the profundal fauna of Loch Awe is very limited, resulting from a single dredge sample taken in 50 m of water. *Tubifex tubifex* was the only oligochaete collected. A variety of Chironomidae were identified (7 genera) consisting of; the tanypod *Procladius*, Chironominae *Cryptochironomus, Endochironomus, Tanytarsus* types A and B, *Polypedilum* and finally the orthoclad *Metriocnemus*. No bivalves were collected in this sample. However, samples taken from the profundal of the Cruachan pumped-storage reservoir in 1979 (which operates by pumping water from Loch Awe) did contain *Pisidium personatum* (Smith 1980). As the Cruachan Reservoir is an artificially constructed water body it seems reasonable to suppose that this bivalve had colonised the reservoir from Loch Awe or its tributaries.

Loch Ness

Samples of profundal fauna were not collected from Loch Ness during the present survey. The information presented in Table 8.2 is taken from the bathymetrical survey of Scottish fresh waters undertaken by Murray & Pullar (1910). This list of the profundal fauna is the result of only one or two dredgings taken at depths of 90–230 m. The only bivalve collected was *Pisidium pusillum* (Gmelin) but it is not possible to say definitely which species this is under present nomenclature, as authors prior to 1917 applied this name to a variety of species. However, material collected from Loch Ness in 120 m of water in 1903 and identified as *P. pusillum* was re-examined by Oldham (1932). This examination revealed two species of *Pisidium* which were identified as *P. personatum* and *P. conventus*. Oligochaetes were represented by the lumbriculid, *Stylodrilus heringianus*, and an unidentified species.

214

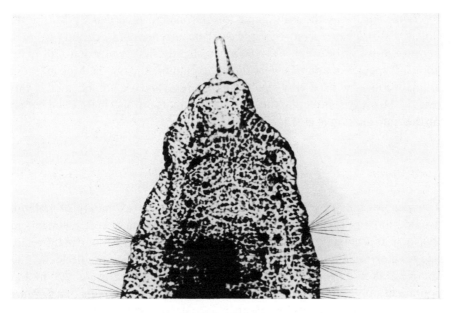

Plate 8.2 Arcteonais lomondi, a species of worm first described from Loch Lomond by Martin (1907) who named it accordingly. It is now known to occur also in Loch Morar. (Photo: P. S. Maitland.)

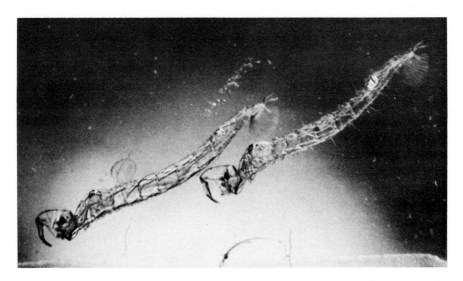

Plate 8.3 Chaoborus flavicans, the larva of which occurs in Loch Lomond but none of the other large lochs, spends part of its time in the profundal muds as a member of the zoobenthos and part in open water as one of the zooplankton. (Photo: D. B. C. Scott.)

215

Chironomus was the only genus of chironomid recorded by Murray & Pullar (1910). They also noted the 'casual' occurrence of *Hydra* and *Lymnaea peregra* in water between 120–180 m. In slightly shallower water (70–90 m) about 40 species of animals were found, excluding Rhizopoda. Most of these animals concerned are very small (e.g. Rotifera) and are not included in this review. However, insect larvae, worms and mites are also mentioned but no further information is given.

Loch Morar

The samples taken by Morgan in Loch Morar at approximately 20 m depth provide most of the information available on the fauna of soft sediments in this loch. It is unlikely that many of these species would be found in the really deep water (maximum depth more than 300 m) of much of the loch. The samples taken during the present survey were collected at *ca.* 200 m and contained only two species of lumbriculid worm – *Lumbriculus variegatus* and *Stylodrilus heringianus*. Neither species was collected in the earlier samples which contained Naididae (including *Arcteonais lomondi*) and Tubificidae. The chironomid *Tanytarsus* and unidentified *Pisidia* were also collected by Morgan.

Loch Shiel

Samples taken during the present survey provide most of the information available on the profundal zoobenthos of Loch Shiel. *Pisidium personatum* represents the bivalve molluscs. Oligochaetes consisted of three species of tubificid, *Limnodrilus hoffmeisteri*, *Tubifex tubifex* and *Peloscolex ferox*. Three species of Chironomidae were also identified – *Metriocnemus* (Orthocladiinae) and two species of *Polypedilum* (Chironominae). These results coincide with the earlier work carried out by Morgan where Tubificidae and Chironomidae are recorded.

Discussion

Though some published data are available on the physico-chemical characteristics of the profundal sediments of several of the large lochs (Murray & Pullar 1910, Gorham 1958, Pennington *et al.*, 1972, etc.) or of other Scottish

216

lochs (Murray & Pullar 1910, Holden 1961, Maitland *et al.* 1981a, etc.), few have examined variation in sediment composition within one loch. Exceptions to this are the work on the sediments of Loch Lomond by Slack (1954) and of Loch Leven by Calvert (1974) and Maitland (1979). Relatively little information too has been published on vertical variation in the composition of the profundal sediments with depth, though some data are available from Loch Lomond (Dickson *et al.* 1978), Loch Ness (Pennington *et al.* 1972), and Loch Morar (C. Muir, personal communication) and other waters, e.g. Kilbirnie Loch (Maitland *et al.* 1981a).

In physical terms the profundal sediments of all the large lochs are characterised by a high proportion of very fine fractions (Table 8.1) – especially clay – and this agrees with the few previous data available from these lochs (Murray & Pullar 1910, Slack 1954). In general, the sediments from very deep water in these lochs are considerably finer than those from much shallower lochs: thus whilst all large lochs examined in this study have sediments with more than 60% of the particles less than 63 μ in size, the equivalent value from the shallow Loch Leven is only 34% (Maitland 1979).

The sediment samples collected by Murray & Pullar (1910) were further analysed chemically by Gorham (1958) who showed that there was no discernible difference between the organic content of muds from small and large or from shallow and deep lochs. For example, though he found that the organic content (expressed by loss on ignition) of profundal sediments from the deepest loch (Morar: 41%) had the fourth highest value, the highest value came from Littlester Loch where mean depth is less than 2 m. The single value of 11.8% (Table 8.1) from Loch Lomond lies within the range found by Slack (1954) there, though the values for Loch Awe (23.9) and Loch Shiel (24.4) are both higher than any analyses from Loch Lomond. The rather low level of organic matter (8.9%) in the sample from Loch Ness fits with the observation by Pennington *et al.* (1972), that the effect of wave action there is considerable and there are virtually no stratified organic deposits down to depths of at least 50 m.

The levels of nitrogen in the profundal sediments are, as might be expected, very closely related to those for loss on ignition (Maitland 1979). The values for phosphorus confirm the suggestion of other studies in this series (e.g. Maitland 1981, Bailey-Watts & Duncan 1981a,b, etc.) that Lochs Lomond and Awe are potentially the richest of the five large lochs studied. Following earlier work by Mortimer (1941) and Holden (1961), Maitland (1978) has shown that sediments from the south basin of Loch Lomond are richer in phosphorus than those from the north basin.

The levels of calcium in the Loch Lomond sediments are surprisingly low, but overall the values for this element, for iron and silicon fall within the ranges known to occur in other British lakes (Gorham 1964, Maitland *et al.* 1981a; etc.). It seems that, for these and other sediment parameters, the range within the large lochs is as great as the differences between them. Certainly a wide range of values, usually well-correlated with depth (cf. Gorham (1958), quoted above), occur in the two lochs which have been intensively studied: Loch Lomond (Slack 1954) and Loch Leven (Maitland 1979).

Apart from the thorough survey undertaken by Slack (1965) the faunistic results presented in this review are rather scanty and variable, taken as they were by a variety of authors using different sampling equipment, in different water depths and on different dates. No doubt the problem of collecting samples from such very deep water has limited the amount of work undertaken in this habitat. However, if minor elements of the faunas are discounted, then each loch has basically a very similar and what may be described as a typical profundal fauna of bivalve molluscs, oligochaete worms and chironomid larvae (Maitland 1978).

The similarity of the zoobenthos communities is probably accounted for by the many features which the profundal habitats of all five lochs have in common. At these great depths there is no autotrophic plant production because of the lack of light and the flora consists only of bacteria and fungi. Water temperatures are always low and vary little seasonally (Ruttner 1963). Substrates consist of very fine sands, silt and clays. The animals that are adapted to these conditions are filter feeders such as *Pisidium*, some Chironominae and Orthocladiinae. Burrowing forms feeding directly on the mud and its microflora and fauna (e.g. oligochaetes) are important also, whilst carnivorous forms such as leeches, Tanypodinae, Chaoborinae and Hydracarina may occur in small numbers.

Various species of *Pisidium* have been recorded from the large lochs. *P. conventus,* found in deep water in Lochs Lomond and Ness is typically an inhabitant of alpine lakes and mountain tarns, where the temperature of the water is always low (Ellis 1940). It is considered to be an ice-age relict species in Loch Lomond (Hunter & Slack 1958). Other cold water species found in the large lochs include the chironomid *Sergentia coracinus.* This species was regarded by Slack (1965) as a characteristic chironomid species of the north basin of Loch Lomond and was almost confined to it. The glacial relict species *Mysis relicta* (recorded from Ennerdale Water in England (Gledhill *et al.* 1976)) was not found in any of the lochs.

Fewer identification problems are encountered with oligochaete worms and for this group, where most animals have at least been tentatively identified, there is a marked similarity among the lochs. *Lumbriculus variegatus*

218

Plate 8.4 Chironomus anthracinus, one of the many species of chironomid midge larvae which occur in the bottom sediments of large lochs. (Photo: W. N. Charles.)

219

and *Stylodrilus heringianus* have been found in the profundal of Lochs Lomond and Morar; the latter species has also been found in Loch Ness. Tubificidae have been collected from all lochs except Loch Ness, which may be accounted for by the low level of organic material in sediments from this loch. *Arcteonais lomondi* is the only naid worm (except *Paranais* sp.) to range beyond the phytal zone of the littoral zone (Weerekoon 1956). It was frequently found at depths of up to 60 m over the whole of Loch Lomond (Slack 1965) and has also been collected in Loch Morar.

Chironomid genera found in two or more lochs are *Procladius, Metriocnemus, Chironomus, Cryptochironomus, Endochironomus, Polypedilum* and *Tanytarsus*. Most of these species are tube dwellers and feed on the fine particulate organic matter falling from the surface waters of the loch. However, *Procladius* and *Cryptochironomus* are both active predatory species not reliant on bottom sediments directly for their food. The *Chironomus* larvae collected from Loch Ness were not identified to species and this genus is more typical of eutrophic lakes. In Loch Lomond the only *Chironomus* larvae found were *C. anthracinus* (Plate 8.4). It appeared in the south basin profundal sediments from September to June as a late instar and Slack (1965) concluded it must have been an immigrant. Only a few were found in each sample but the main source of the population was not known as the species did not generally occur in the littoral zone.

It is clear that much more information is required before adequate species lists can be compiled. The few samples taken in Lochs Awe, Ness, Morar and Shiel provide no information on the abundance, seasonality or productivity of the profundal fauna in these lochs. The work of Slack (1965) clearly indicated regional differences in the distribution of species among the different basins of Lomond. Although markedly different basins do not occur in the other four lochs, spatial differences are likely, especially in Loch Awe. In the other studies in this series (e.g. Smith *et al.* 1981a) it has been possible to grade the lochs in order of the abundance and diversity of their floras and faunas. The extremely limited information available for this review is inadequate to allow an arrangement of the lochs in this way as far as their profundal faunas are concerned.

Acknowledgements

We are grateful to Dr A. E. Bailey-Watts, Ms G. M. Dennis, Mr A. A. Lyle, Mr A. Kirika, Mrs S. M. Pennock and Ms A. E. Sneddon for help with field work. We would like to thank Professor P. G. Jarvis for facilities provided for sediment analysis at the University of Edinburgh. Unpublished data from Lochs Morar and Shiel were provided by Mr N. C. Morgan. Dr H. D. Slack was kind enough to review the manuscript, which was typed by Mrs M. S. Wilson.

References

Allen, S. E., Grimshaw, H. M., Parkinson, J. A. & Quarmby, C., 1974. Chemical analysis of ecological materials. Oxford: Blackwell.

Bailey-Watts, A. E. & Duncan, P., 1981a. The ecology of Scotland's largest lochs: Lomond, Awe, Ness, Morar and Shiel. Ed. P. S. Maitland. Chemical characterisation – a one year comparative study. Chap. 3. This volume.

Bailey-Watts, A. E. & Duncan, P., 1981b. The ecology of Scotland's largest lochs: Lomond, Awe, Ness, Morar and Shiel. Ed. P. S. Maitland. The phytoplankton. Chap. 4. This volume.

Bailey-Watts, A. E. & Duncan, P., 1981c. The ecology of Scotland's largest lochs: Lomond, Awe, Ness, Morar and Shiel. Ed. P. S. Maitland. A review of macrophyte studies. Chap. 5. This volume.

Calvert, S. E., 1974. The distribution of bottom sediments in Loch Leven, Kinross, Proc. R. Soc. Edinb. B. 74: 69–80.

Crooke, W. M. & Simpson, W. E., 1971. Determination of ammonium in Kjeldahl digests of crops by automated procedure. J. Sci. Fd Agric. 22: 9–10.

Cummins, K. W., 1962. An evaluation of some techniques for the collection and analysis of benthic samples with special emphasis on lotic waters. Am. Midl. Nat. 67: 477–504.

Dickson, J. H., Stewart, D. A., Thompson, R., Turner, G., Baxter, M. S., Drndarsky, N. D. & Rose, J., 1978 Palynology, palaeomagnetism and radiometric dating of Flandrian Marine and freshwater sediments of Loch Lomond. Nature, Lond., 274: 548–553.

Ellis, A. E., 1940. The identification of British species of *Pisidium*. Proc. malac. Soc. Lond. 24: 44–88.

Fedoruk, A. M., 1964. Studies on the biology, ecology, and dynamics of profundal Chironomidae in Loch Lomond. Ph.D. Thesis, University of Glasgow.

Gledhill, T., Sutcliffe, D. W. & Williams, W. D., 1976. Key to British freshwater Crustacea: Malacostraca. Freshwater Biological Association Scientific Publication, No. 32.

Gorham, E. V., 1958. The physical limnology of northern Britain: an epitome of the bathymetrical survey of the Scottish freshwater lochs, 1897–1909. Limnol. Oceanogr. 3: 40–50.

Gorham, E. V., 1964. Molybdenum, manganese and iron in lake muds. Verh. int. Verein. theor. angew. Limnol. 15: 330–332.

Holden, A. V., 1961. The removal of dissolved phosphate from lake waters by bottom deposits. Verh. int. Verein. theor. angew. Limnol. 14: 247–251.

Hunter, W. R. & Slack, H. D., 1958. *Pisidium conventus*. Clessin in Loch Lomond. J. Conch. 24: 245–247.

Maitland, P. S., 1978. Biology of fresh waters. Glasgow: Blackie.

Maitland, P. S., 1979. The distribution of sediments and zoobenthos in Loch Leven, Scotland. Arch. Hydrobiol. 85: 98–125.

Maitland, P. S., 1981. The ecology of Scotland's largest lochs: Lomond, Awe, Ness, Morar and Shiel. Ed. P. S. Maitland. 1. Introduction and catchment analysis. Chap. 1. This volume.

Maitland, P. S., Smith, I. R., Bailey-Watts, A. E., East, K., Morris, K. H., Lyle, A. A. & Kirika, A., 1981a. Kilbirnie Loch: an ecological assessment. Glasg. Nat. 20: 7–23.

Maitland, P. S., Smith, B. D. & Dennis, G. M., 1981b. The ecology of Scotland's largest lochs: Lomond, Awe, Ness, Morar and Shiel. Ed. P. S. Maitland. The crustacean zooplankton. Chap. 6. This volume.

Martin, C. H., 1907. Notes on some oligochaetes found on the Scottish Loch Survey. Proc. R. Soc. Edinb. 28: 21–27.

Mortimer, C. H., 1941. The exchange of dissolved substances between mud and water in lakes. J. Ecol. 29: 280–329.

Murphy, J. & Riley, J. P., 1962. A modified single solution method for the determination of phosphates in natural waters. Analytica chim. Acta., 27: 31–36.

Murray, J. & Pullar, L., 1910. Bathymetrical survey of the fresh water lochs of Scotland. Edinburgh: Challenger. Vols. 1–6.

Oldham, C., 1932. Notes on some Scottish and Shetland *Pisidia*. J. Conch. 19: 271–278.

Pennington, W., Haworth, E. Y., Bonny, A. P. & Lishman, J. P., 1972. Lake sediments in northern Scotland. Phil. Trans. R. Soc. Lond. B. 264: 191–294.

Piper, C. S., 1950. Soil and plant analysis. Adelaide: Waite Agricultural Research Institute Monograph.

Ratcliffe, D. A., 1977. A nature conservation review. Cambridge University Press. Vols. 1–2.

Ruttner, F., 1963. Fundamentals of limnology. 3rd edition. Transl. by Frey, D. G. and Fry, F. E. J. University of Toronto Press.

Slack, H. D., 1954. The bottom deposits of Loch Lomond. Proc. R. Soc. Edinb. 65: 213–238.

Slack, H. D., 1965. The profundal fauna of Loch Lomond, Scotland. Proc. R. Soc. Edinb. 69: 272–297.

Smith, B. D., 1980. Further studies on the impact of the proposed Craigroyston pumped-storage hydro-electric scheme on the ecology of Loch Lomond. Unpublished report by the Institute of Terrestrial Ecology to the North of Scotland Hydro-Electric Board.

Smith, B. D., Maitland, P. S., Young, M. R. & Carr, M. J., 1981a. The ecology of Scotland's largest lochs: Lomond, Awe, Ness, Morar and Shiel. Ed. P. S. Maitland. The littoral zoo-benthos. Chap. 7. This volume.

Smith, I. R., Lyle, A. A. & Rosie, A. J., 1981b. The ecology of Scotland's largest lochs: Lomond, Awe, Ness, Morar and Shiel. Ed. P. S. Maitland. Comparative physical limnology. Chap. 2. This volume.

Weerekoon, A. C. J., 1956. Studies on the biology of Loch Lomond. 1. The benthos of Auchentullich Bay. Ceylon J. Sci. 7: 1–94.

Welch, P. S., 1948. Limnological methods. New York: Blakiston.

Wentworth, C. K., 1922. A scale of grade and class terms for clastic sediments. J. Geol. 30: 377–392.

9.   The fish and fisheries

P. S. Maitland, B. D. Smith & S. M. Adair

Abstract

This account of the fish populations of Scotland's largest lochs is based on unpublished data available from netting, echo sounding, counting at fish passes and angling. Some of the records are very recent (e.g. from gill netting in Loch Lomond 1978–79) others extend back over 100 years (seine netting in Loch Awe). The data provide a basic background to the fish and fisheries of these large lochs and comparisons among them. It is very clear that Loch Lomond has the most diverse fish community with 15 species, and Lochs Morar and Shiel the least with only five species. The fisheries of Lochs Lomond, Awe and Ness are also much larger and more varied than those of Lochs Morar and Shiel. The reasons for these differences are discussed in relation to natural differences among the five large lochs and to changes due to human pressures.

Introduction

This paper is one of a series describing a recent multidisciplinary study of five of Scotland's largest lochs: Lomond, Awe, Ness, Morar and Shiel. Other papers in the series deal with general aspects of these lochs and their catchments (Maitland 1981), their physical limnology (Smith *et al.* 1981b), chemistry, phytoplankton and macrophytes (Bailey-Watts & Duncan 1981a,b,c), zooplankton (Maitland *et al.* 1981) and zoobenthos (Smith *et al.* 1981a).

The importance of Scotland's large freshwater lochs, though often appreciated aesthetically, is often forgotten as a national resource. Apart from their value to tourism and as a source of water, hydro-power, etc. these waters have a major value to angling in this country. General aspects of the fish

Monographiae Biologicae, Vol. 44, ed. by P. S. Maitland

communities in forty-one of Scotland's largest lochs have been reviewed by Maitland (1976). Five of the largest of these waters are the subject of this paper. The fish populations in most of them are under considerable pressure and have never been adequately studied. Their general scientific merit too is often given insufficient priority.

The study arose following a proposal by the North of Scotland Hydro-Electric Board to build a large pumped-storage hydro-electric power station at Craigroyston on Loch Lomond. Only two such stations have been constructed in Scotland to date – at Cruachan on Loch Awe and at Foyers on Loch Ness. In view of the potential importance of such schemes to the ecology of the lochs concerned, it was decided to carry out a comparative study of the five lochs noted above. The vulnerability of young fish in particular to entrainment in power station intakes (Marcy 1973) emphasises the importance of understanding the nature of the fish communities in these lochs.

Apart from Loch Lomond where several reviews of the status of the fish community (Brown 1891, Lumsden & Brown 1895, Lamond 1931, Hunter *et al*. 1959, Maitland 1972a) and studies of individual fish species (Parnell 1838, McNiven 1863, Robertson 1870, 1888, Lamond 1922, Nall 1933, Gervers 1954, Slack 1955, 1957, Copland 1956, MacDonald 1959, Maitland 1967a, 1969, Mills 1969, Shafi 1969, Shafi & Maitland 1971a, 1971b, Scott 1975, Fuller *et al*. 1976, Maitland 1979) have been carried out, there is little published information about the five large lochs. Several general works, including angling guides (e.g. Calderwood 1921, Stirling 1931, Castle 1937, etc.)

Table 9.1 Basic physical and chemical characteristics of relevance to the fish and fisheries of Lochs Lomond, Awe, Ness, Morar and Shiel

Feature	Lomond	Awe	Ness	Morar	Shiel
Surface area (km^2)	71.1	38.5	56.4	26.7	19.6
Catchment area (km^2)	781	840	1775	168	248
Surface altitude (m)	7.9	36.2	15.8	10.1	4.5
Mean depth (m)	37.0	32.0	132.0	86.6	40.5
Maximum depth (m)	189.9	93.6	229.8	310.0	128.0
Mean discharge (cumecs)	44.5	57.4	81.5	10.7	17.8
Catchment: mean slope (m/km)	180	198	176	303	380
Catchment: % arable ground	17.0	1.6	1.9	0.0	1.4
Catchment: number of lochs	76	156	355	73	60
pH (annual mean)	6.78	6.90	6.70	6.63	6.14
Alkalinity (mg/l CaCO$_3$)	6.37	8.97	4.48	3.27	2.10
Conductivity (K$_{20}$ μS cm^{-1})	34	41	30	35	29

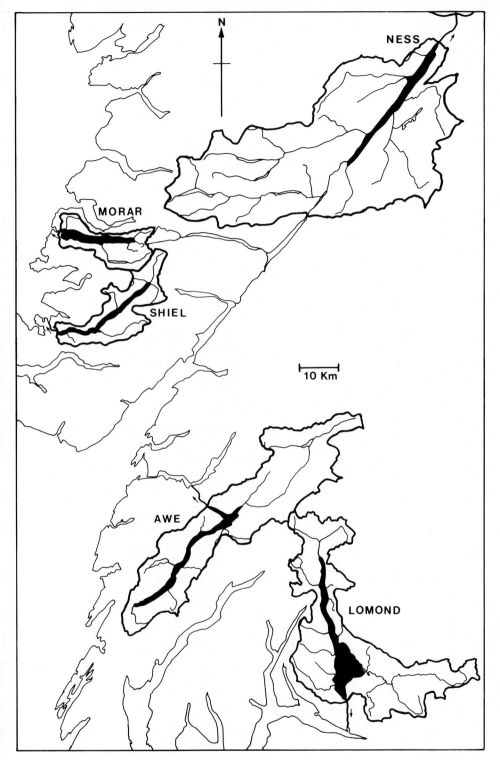

Fig. 9.1 The geographic position in Scotland of Lochs Lomond. Awe, Ness, Morar and Shiel, showing the main inflows, outflows and catchment boundaries.

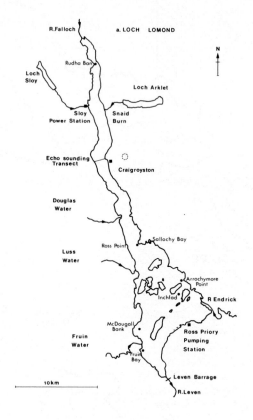

a. LOCH LOMOND

R.Falloch

N

Rudha Ban

Loch
Sloy

Loch Arklet

Sloy
Power Station

Snaid
Burn

Echo sounding
Transect

Craigroyston

Douglas
Water

Luss
Water

Ross Point

Sallochy Bay

Arrochymore
Point

Inchfad

R Endrick

McDougall
Bank

Ross Priory
Pumping
Station

Fruin
Water

Fruin
Bay

Leven Barrage

R.Leven

10km

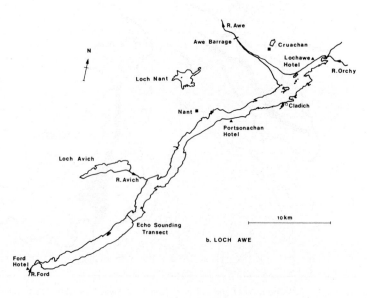

R. Awe

Awe Barrage

Cruachan

N

Lochawe
Hotel

Loch Nant

R.Orchy

Nant

Cladich

Portsonachan
Hotel

Loch Avich

R. Avich

Echo Sounding
Transect

10km

b. LOCH AWE

Ford
Hotel

R.Ford

226

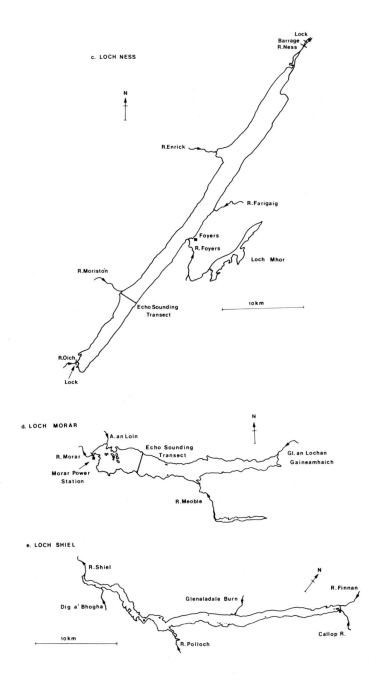

Fig. 9.2 Outline diagrams of the large lochs showing the principal features relevant to this paper: (a) Loch Lomond, (b) Loch Awe, (c) Loch Ness, (d) Loch Morar and (e) Loch Shiel.

227

deal superficially with the larger species present, but apart from short papers such as the brief account of the age and growth of trout and eels in Loch Morar (Anonymous 1972), virtually nothing is known of the ecology of the fish in Lochs Awe, Ness, Morar or Shiel.

The five lochs (Fig. 9.1) have been described by Murray & Pullar (1910) and in detail by Maitland (1981) and Smith et al. (1981b). Only a brief description of their major features is given here. The characters of most relevance to the fish and fisheries are compared in Table 9.1. Loch Lomond (Fig. 9.2a) is the largest area of fresh water in Great Britain and is quite unique in the dual character created by the crossing of the Highland Boundary Fault Line, giving a deep narrow basin to the north and a broad shallow one to the south. Loch Awe (Fig. 9.2b) is quite the longest loch in Great Britain and is also extremely narrow for its length. Its floor includes five separate basins each greater than 60 m in depth. Possibly the claim of Scotland's largest loch must go to Loch Ness (Fig. 9.2c) which holds almost three times the amount of water contained in the next largest by volume (Loch Lomond) and more water than the combined total of all the lakes and reservoirs in England and Wales (Smith & Lyle 1979). Loch Morar (Fig. 9.2d) is very clearly the deepest loch in the entire British Isles with a single main basin running down to a depth of over 300 m. The nearest water deeper than this is in the Atlantic Ocean west of St Kilda. Only six other lakes in Europe are as deep as Loch Morar. Apart from Lochs Lomond, Awe and Ness no other lake in Great Britain is longer than Loch Shiel (Fig. 9.2e); its basin is very narrow, like Loch Awe, and its water regime is probably the most natural of all the lochs considered here.

The status of freshwater fish in Scotland has been reviewed recently by Maitland (1977). Of the British list of 54 species (Maitland 1972b), some 40 are known to occur in Scotland, the remainder being restricted to England, particularly the south-east. Of the 40 Scottish species, nine are essentially brackish and come into fresh water only occasionally or for short periods only. At least fourteen of the remaining species are known to have been introduced to Scotland by Man. None of these occurs in the five large lochs studied here.

228

Methods and information

General

Detailed research on fish populations is notoriously difficult and demanding of manpower and other resources. The logistics and biological problems of such work are even more complex when dealing with the fish communities of the large lochs studied here, and indirect methods are often the only way to obtain any insight. This paper is a synopsis of the information available to date on the fish and fisheries of Lochs Lomond, Awe, Ness, Morar and Shiel, and includes data from numerous nettings, from echo soundings, from fish counters and from angling returns. Some of the species are shown in Plates 9.1-9.

Netting

Most of the information obtained by netting is from Loch Lomond where, in addition to commercial net fisheries which have operated for over a hundred years (Lamond 1931), netting for research purpose (using seine, gill and trap nets) has been carried out at various times mainly under the direction of Dr H. D. Slack and one of the authors (PSM) since 1950. Much of this research has concerned the powan, *Coregonus lavaretus* (Slack 1957, Maitland 1969) and most netting done in the south basin of the loch. However, some data are available from recent nettings in the northern basin. The principal sites where netting has been carried out are indicated in Fig. 9.2a.

As part of a multidisciplinary project to assess the impact of the proposed pumped-storage hydro-electric project at Craigroyston on the ecology of Loch Lomond, regular netting was carried out in the small bay just beside the site of the proposed power station (Fig. 9.2a). Here, a standard programme of gill netting was carried out on eight occasions from March 1978 to January 1979. Each time, eight multifilament nylon gill nets were set as soon as possible after noon on the first day in regular positions in the bay at Craigroyston (Fig. 9.3) and lifted before noon on the second day. Two of the nets were of a single very fine mesh (5 and 8 mm bar) and 50 m in length. The remaining six nets were identical to each other and manufactured in Sweden especially for this work. Each was 50 m long and 2 m deep with 10 panels each of a different mesh size, ranging from 10 to 50 mm bar. All fish caught in the nets were identified, measured for length and counted.

Apart from data from Loch Awe where a private seine net fishery has been

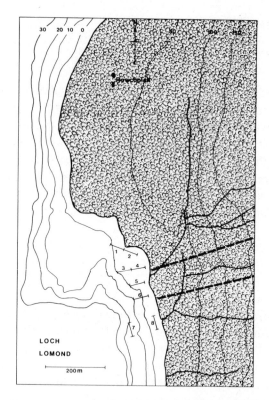

Fig. 9.3 Outline diagram of the Craigroyston area of Loch Lomond showing the positions of standard gill nets fished during 1978–79, in relation to topography and the intake/discharge pipes of the proposed power station – these are indicated by the heavy dashed lines. The contours are heights above and below sea level.

operating in the Cladich area for over 100 years, there is little netting information available from the other large lochs. In conjunction with the echo sounding described below, standard gill netting was carried out in 1966 in Lochs Lomond, Awe, Ness and Morar, but not in Loch Shiel. Occasional netting has also been carried out from time to time by other research workers in Lochs Awe, Ness and Morar, but again apparently never in Loch Shiel.

Echo sounding

The use of echo sounding equipment is one of the best ways to overcome some of the problems of studying fish populations in large bodies of water. Equipment obtained in 1966 and used very successfully on Loch Lomond was

230

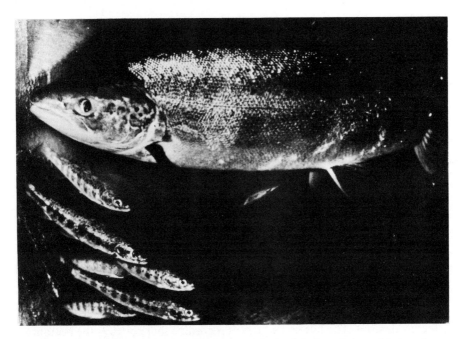

Plate 9.1 Salmon, *Salmo salar:* the major sport species in all the large lochs and their outflows. (Photo: North of Scotland Hydro-Electric Board.)

Plate 9.2 Trout, *Salmo trutta:* both sea trout and brown trout are common in all the large lochs. (Photo: K. H. Morris.)

used in August and September of that year to obtain some preliminary information on fish in several other Scottish lochs including Awe, Ness and Morar – as well as Loch Lubnaig and the Mill Loch, Lochmaben (Maitland 1967b). The technique in carrying out these soundings was the same at each loch: at each a transect was established at a suitable locality (Fig. 9.2a-2d) and soundings were taken along this transect line every 3 h over a 24 h period. Between these transect runs, stationary soundings were taken for about 40 min at a buoy moored in deep water near the transect line. Prior to the soundings being taken, gill nets of varied mesh sizes were set near by at various depths in the water from bottom to surface. These were examined at intervals between soundings.

The machine used for these soundings is a portable straight line recording echo sounder which gives a continuous record on dry recording paper. It has two scale ranges, one of 0-110 m phasing continuous to 340 m, and the other of 0-220 m phasing continuous to 680 m. The transmission rate at these ranges is 120 and 60 soundings/min respectively. The transmission system is based on a pulsed c.v. transistor transmitter with an operating frequency of 48 kc/s. This machine is capable of producing both a normal sounding, and a white line effect as desired. The latter is a form of recording that portrays the bottom as a thin dark line with a white band immediately below it. White line recording makes it much easier to discern substrate echoes from echoes of fish close to the bottom.

Counting at fish passes

As part of their programme to conserve fish stocks, especially salmon, the North of Scotland Hydro-Electric Board (1979) have installed fish passes or Borland fish locks beside many of their dams. Associated with these in many places are electronic fish counters which can distinguish between ascending and descending fish. The counters together with visual counts at some of the Borland passes are used to provide data on the movement of large migratory fish. Three of the lochs concerned in this study (Awe, Ness and Morar) have fish passes and counters associated with them and the data have been made available to the authors.

At Loch Awe (Fig. 9.2b), a barrage across the river as it leaves the loch diverts water into a 5 km long tunnel to Inverawe power station near the mouth of the river. The Awe barrage raised the water in a short stretch of the river channel to the same level as loch Awe without altering the water level of the loch itself. Incorporated in the barrage is a Borland fish lift (records

232

available since 1964), while a fish counter operates at Inverawe power station (installed 1966). At Loch Ness (Fig. 9.2c) there is no information on fish entering the loch itself, but two of its major tributaries have fish passes. A small dam on the River Moriston forms Loch Dundreggan, and here a Borland fish pass has been provided (records available from 1969) so that fish can continue upstream as far as Ceannacroc. On the River Garry (whose waters run into Loch Oich and then into Loch Ness) a dam has raised the level of the original Loch Garry and includes a Borland type fish pass (records available from 1966). At Loch Morar (Fig. 9.2d) a small dam was constructed at the exit in 1948 to provide hydro-electricity. Subsequently a fish counter was installed in the pass associated with this dam and this was brought into operation in 1966.

Angling records

All five of the lochs discussed here have important rod fisheries for salmon and trout and certain data are available from these. At Loch Lomond the first attempt at organised angling started about 1858 and by the end of the century the present Loch Lomond Angling Improvement Association was fully established. This has kept regular records since about 1900 of its membership and of monthly catches of salmon, sea trout and brown trout in Loch Lomond and its tributaries. These have been reviewed by Maitland (1972a) and are brought up to date in the present paper.

At Loch Awe several of the old-established fishing hotels (Fig. 9.2b) have continuous runs of fish catches and useful data are available from Lochawe Hotel (no longer operating) at the north end (1887–1909), Ford Hotel at the south end (1926–1966) and Portsonachan Hotel about the middle of the loch (1931–1977). In addition, a shore seine net fishery has operated at Cladich at the north end from 1871 to the present. The catches from all these fisheries have been made available to the authors.

Annual returns of the catches of salmon and sea trout must be made to the Secretary of State for Scotland (Department of Agriculture and Fisheries) for each fishery by its owner or his lessee. The rod returns for the Clyde & Leven (Lomond), Awe, Ness, Morar and Shiel districts are composed largely of catches from the relevant lochs or their tributaries, and data from 1952 to 1978 have been made available to the authors. There is also scattered information on fish in these lochs in the literature (e.g. Wood 1954, Campbell 1979). Unfortunately the official statistics are confidential and only relative values can be discussed here.

Results

General

The basic information from the above sources is shown in Fig. 9.4-9 and Tables 9.2-7. In general the data are presented according to the method used

Table 9.2   The occurrence of fish species in Lochs Lomond, Awe, Ness, Morar and Shiel

Fish species	Lomond	Awe	Ness	Morar	Shiel
Sea Lamprey					
Petromyzon marinus Linnaeus 1758	+	–	–	–	–
River Lamprey					
Lampetra fluviatilis (Linnaeus 1758)	+	–	–	–	–
Brook Lamprey					
Lampetra planeri (Bloch 1784)	+	–	+	–	–
Salmon					
Salmo salar Linnaeus 1758	+	+	+	+	+
Trout					
Salmo trutta Linnaeus 1758	+	+	+	+	+
Charr					
Salvelinus alpinus (Linnaeus 1758)	–	+	+	+	–
Powan					
Coregonus lavaretus (Linnaeus 1758)	+	–	–	–	–
Pike					
Esox lucius Linnaeus 1758	+	+	+	–	–
Roach					
Rutilus rutilus (Linnaeus 1758)	+	–	–	–	–
Minnow					
Phoxinus phoxinus (Linnaeus 1758)	+	+	–	–	–
Stone Loach					
Noemacheilus barbatulus (Linnaeus 1758)	+	–	–	–	–
Eel					
Anguilla anguilla (Linnaeus 1758)	+	+	+	+	+
Three-spined Stickleback					
Gasterosteus aculeatus Linnaeus 1758	+	+	+	+	+
Ten-spined Stickleback					
Pungitius pungitius (Linnaeus 1758)	+	–	–	–	–
Perch					
Perca fluviatilis Linnaeus 1758	+	+	–	–	–
Flounder					
Platichthys flesus (Linnaeus 1758)	+	–	–	–	+
Total number of species	15	8	7	5	5

234

Plate 9.3 Charr, *Salvelinus alpinus:* found in Lochs Awe, Ness and Morar, but not in Loch Lomond nor, apparently, in Loch Shiel (Photo: G. F. Friend.)

Plate 9.4 Powan, *Coregonus lavaretus:* abundant in Loch Lomond, but occurs in none of the other large lochs. It is found elsewhere in Scotland only in Loch Eck (Photo: R. J. Roberts.)

Plate 9.5 Minnow, *Phoxinus phoxinus:* occurs in Lochs Lomond and Awe, but apparently not yet in Lochs Ness, Morar and Shiel. (Photo: K. H. Morris.)

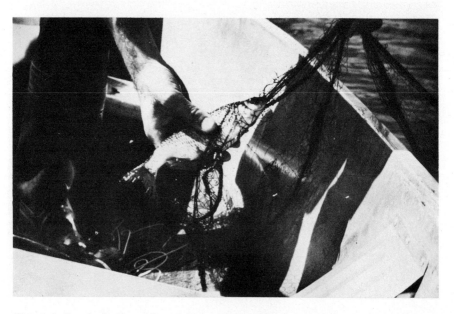

Plate 9.6 Roach, *Rutilus rutilus:* of the large lochs, this species occurs only in Loch Lomond, its most northerly habitat on the west coast of Great Britain. (Photo: K. R. Maitland.)

236

and, where possible, in a comparative way among the five lochs. However, because of the variation of the level of information from each loch, the results are first considered here on a loch by loch basis, before comparative aspects are considered (see discussion).

Loch Lomond

It can be seen from Table 9.2 that fifteen species of fish occur in Loch Lomond, notable among which is the powan, *Coregonus lavaretus*. For the five small species which are found here (brook lamprey, minnow, stone loach, three-spined stickleback and ten-spined stickleback), because the methods used were appropriate for larger fish only, relatively little information is available from the present study. However, other work (MacDonald 1959, Maitland 1965) has shown that they are all common in the loch (in the south end at least) or its tributaries. The same is true to a certain extent of sea lamprey, river lamprey and eel, all of which are rarely taken by the methods used here. Though marine lampreys are apparently uncommon, river lampreys are abundant in the loch and parasitize powan there heavily, as has been demonstrated by a recent study (Maitland 1979). Eels are very common too, they frequently attack other fish caught in gill nets and have formed the basis of an intermittant fishery from time to time in the past.

All methods of capture used here are selective among various species and seasonally within a species and the results must be interpreted accordingly. Of the larger species (Table 9.3) trout and powan appear to be the most abundant, with roach and perch quite numerous also. Salmon and flounder

Table 9.3a   Data from 106 seine net hauls in Loch Lomond: 1951–58

Netting data	Salmon	Trout	Powan	Pike	Roach	Perch	Flounder	Others	Overall
Total number caught	11	205	2955	40	251	180	12	4	3658
% occurrence in hauls	8	43	74	14	13	16	9	2	100
Mean number per haul	0.10	1.93	27.88	0.38	2.37	1.70	0.11	0.04	34.51
Minimum number per haul	0	0	0	0	0	0	0	0	0
Maximum number per haul	2	20	341	10	100	114	2	3	342

237

Table 9.3b   Data from 38 gill nettings in Loch Lomond in 1954–58, and 10 in 1964–66

Netting data	Salmon	Trout	Powan	Pike	Roach	Perch	Flounder	Others	Overall
1951–58									
Total number									
caught	6	60	1317	3	56	5	1	0	1448
% occurrence									
in hauls	13	63	87	8	53	8	3	0	100
Mean number									
per haul	0.16	1.58	34.66	0.08	1.47	0.13	0.03	0	38.11
Minimum number									
per haul	0	0	0	0	0	0	0	0	0
Maximum number									
per haul	2	6	128	1	8	3	1	0	129
1964–66									
Total number									
caught	1	162	599	7	36	7	1	0	813
% occurrence									
in hauls	10	100	100	40	30	40	10	0	100
Mean number									
per haul	0.10	16.20	59.90	0.70	3.60	0.70	0.10	0	81.30
Minimum number									
per haul	0	3	5	0	0	0	0	0	29
Maximum number									
per haul	1	40	262	3	21	3	1	0	267

appear to be much less common than the other species, but this may be because the former inhabits deeper water off-shore, while the latter is almost entirely bottom-dwelling. Both would be less vulnerable than the other species to the nets used here. All fish species appear to occur throughout the loch but pike and roach are much more common in the southern basin than in the northern one.

The most extensive netting programme was that carried out at Craigroyston in the inshore area beside the intake/discharge point of the proposed pumped-storage hydro-electric station. The general results (Table 9.4) show that trout, powan and perch were the common species here at the time of the study, though salmon, roach, minnows and eel occurred also (cf. Table 9.5). Considerable seasonal differences were apparent (Fig. 9.4a) with trout most common during the winter period but powan and perch during the summer. Length frequency analyses indicate the sizes of fish present at different times of the year (Fig. 9.4b). These are relevant to their vulnerability to being taken in by the pumping system of the proposed station.

238

Table 9.4 Combined fish catch from 8 standard gill nets set at Craigroyston, Loch Lomond, 1978–79. On some dates, eels were known to be present (+) but not caught

	8.3.78	20.5.78	27.6.78	25.7.78	23.9.78	28.10.78	2.12.78	13.1.79	Total
Salmon	—	—	—	1	1	—	—	—	2
Trout	59	20	5	2	9	26	55	50	226
Powan	—	2	30	33	31	16	9	18	139
Roach	—	—	—	—	1	—	—	—	1
Minnow	1	2	—	—	—	6	1	—	10
Eel	—	—	+	+	+	+	+	1	1
Perch	1	43	52	84	41	10	8	1	240
Total	61	67	87	120	83	58	73	70	619

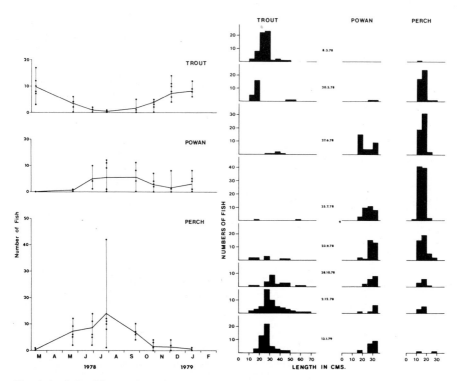

Fig. 9.4a (left) The number of the commonest fish (trout, powan and perch) caught in 6 standard gill nets fished in the Craigroyston area of Loch Lomond during 1978–79. The points on vertical bars represent the actual numbers in each net; the lines join the mean numbers caught on each date.

Fig. 9.4b (right) The length-frequencies of trout, powan and perch caught in 6 standard gill nets fished in the Craigroyston area of Loch Lomond during 1978–79.

239

Table 9.5 Numbers of fish caught in occasional gill nets set in Loch Awe, Ness and Morar

Netting	Salmon	Trout	Charr	Pike	Perch	Others	Total
Ness (11.9.66)	1	20	13	0	0	0	34
Morar (13.9.66)	0	23	8	0	0	0	31
Awe (22.9.66)	0	29	0	0	0	0	29

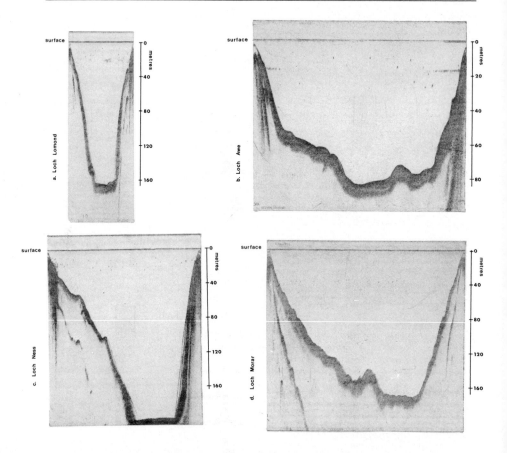

Fig. 9.5 Typical daylight echo soundings from transects across (a) Loch Lomond, (b) Loch Awe, (c) Loch Ness and (d) Loch Morar. The dots in the surface waters represent individual fish.

Echo soundings taken in both the northern (Fig. 9.5a) and southern basins of Loch Lomond indicate that fish are restricted almost entirely to the littoral area or to the upper layers in open water. It is very rare to detect fish below depths of 50m on echo soundings and none have ever been caught in nets set

240

Table 9.6 Counts of fish in inshore and offshore areas of echo sounding transects from Lochs Lomond, Awe, Ness and Morar. The numbers are fish per km of transect with mean (minimum and maximum) values from six independent counts in each area. All soundings were taken during daylight in August 1966

	Lomond	Awe	Ness	Morar
Offshore	36 (22–55)	68 (29–98)	40 (25–50)	39 (30–49)
Inshore	46 (31–71)	57 (18–91)	56 (33–85)	30 (12–49)
Total transect	41 (22–71)	63 (18–98)	48 (25–85)	35 (12–49)

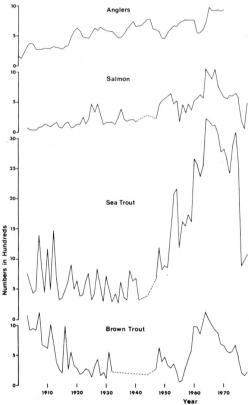

Fig. 9.6 Basic data from the records of the Loch Lomond Angling Improvement Association: 1900–78.

in water deeper than this. Counts of fish in echo transects are given in Table 9.6.

The data available from the sport fishery for salmon and trout cover the period from 1900 to 1978 (Fig. 9.6). The annual catches during this period vary

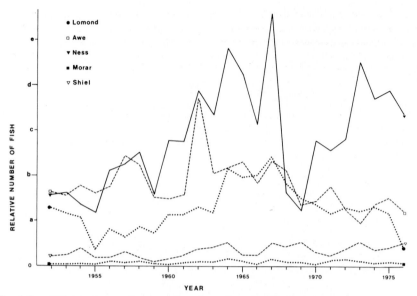

Fig. 9.7a Official returns of rod-caught salmon from the fishery districts of Clyde & Leven (Lomond), Awe, Ness, Morar and Shiel: 1952–76. The values indicated are relative ones only, for the actual data are confidential and not yet available for publication.

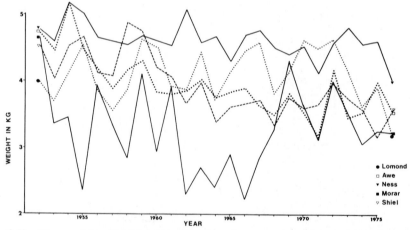

Fig. 9.7b Mean weight of rod-caught salmon from the fishery districts of Clyde & Leven (Lomond), Awe, Ness, Morar and Shiel: 1952–76.

considerably from year to year but there is little evidence of any specific pattern other than a distinct increase in catches over the last 30 years. This coincides with and is almost certainly related to a parallel rise in the number

242

of people fishing, as indicated by the membership of the Loch Lomond Angling Improvement Association. Several species of fish have been introduced to Loch Lomond (e.g. tench, *Tinca tinca* (L.)) but none have ever become established (Young 1870, Maitland 1972a).

Loch Awe

The fish community at Loch Awe consists of eight species (Table 9.2), five of which are either migratory or have marine affinities and are likely to have been there for several thousand years (Maitland 1977). Only three (pike, minnow and perch) are purely freshwater forms and likely to have been there for a much shorter period. All the species appear to be common in at least some part of the loch. As in Loch Lomond, virtually all fish are found in the upper waters and virtually never below 40 m (Fig. 9.5b).

The annual counts of salmon running into Loch Awe, (Fig. 9.8) though they fluctuate from year to year, appear to have a reasonably consistent pattern

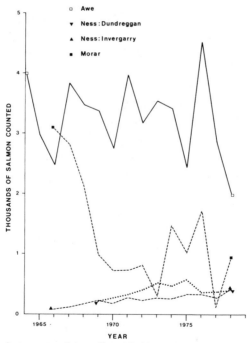

Fig. 9.8 Numbers of salmon counted ascending to Loch Awe and Loch Morar, and from Loch Ness (Dundreggan and Invergarry).

243

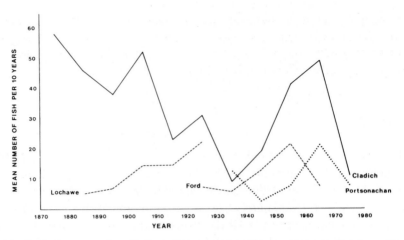

Fig. 9.9 Basic data on catches of salmon (5-y means) from three hotel records (Lochawe, Ford and Portsonachan) and one net fishery (Cladich) on Loch Awe.

around a mean of 3247 fish/y. The highest value since records began was 4516 in 1976 and the lowest 2000 in 1978.

The records from the seine net fishery at Cladich and the fishing hotels at Ford, Portsonachan and Lochawe (Fig. 9.9) though difficult to interpret because of lack of information on fishing effort, do give some idea of the number of salmon taken by thse fisheries over a period of more than 100 years. The total numbers caught by rod in the Awe district show annual fluctuations (Fig. 9.7a) and a downward trend over the period 1964–1976. The mean number of fish taken annually over this period in the Awe Fishery District is a relatively high fraction of the number recorded as entering Loch Awe at this time. If these figures are realistic, then a very significant proportion of the population is killed by anglers before spawning. During 1952–76 the mean weight of salmon caught in the district has dropped markedly (Fig. 9.7b).

Loch Ness

Of the limited community of seven species of fish which occur in Loch Ness (Table 9.2), only one (pike) does not have marine affinities and presumably is the most recent invader. All seven species appear to be common in at least some parts of the loch. As in the other lochs, fish occur only in the surface waters (Fig. 9.5c) and are rarely found here below a depth of about 30 m.

The annual runs of salmon ascending from Loch Ness into the Garry and

244

the Moriston to pass the counters at Invergarry and Dundreggan respectively are relatively low and consistent from year to year (Fig. 9.8), though with a definite upward trend over the period concerned. The rod returns for the Ness district (Fig. 9.7a) are much more variable from year to year with notable maxima in 1964 and 1967 and minima in 1955 and 1969 respectively. As at Loch Awe, there has been a steady decline in the mean weight of salmon over the last 25 years (Fig. 9.7b).

Loch Morar

Loch Morar has a fish community consisting of only five species – all with marine affinities (Table 9.2). All of these fish seem reasonably common there. As in the other large lochs, most fish occur in the upper waters (Fig. 9.5d) and rarely below about 45 m.

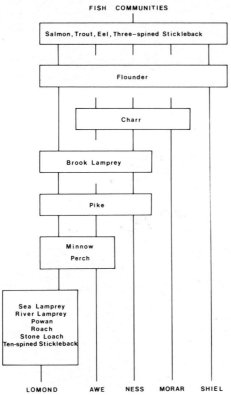

Fig. 9.10 Species associations in the fish communities of Lochs Lomond, Awe, Ness, Morar and Shiel.

245

Plate 9.7   Three-spined Stickleback, – *Gasterosteus aculeatus:* one of the few species which occur in all the large lochs. (Photo: K. H. Morris.)

Plate 9.8   ⎱Ten-spined Stickleback, *Pungitius pungitius:* common in Loch Lomond but found in none of the other large lochs. (Photo: K. H. Morris.)

Plate 9.9   Perch, *Perca fluviatilis:* found in Lochs Lomond and Awe, but none of the other large lochs. (Photo: K. H. Morris.)

The counts of salmon running into Loch Morar (Fig. 9.8) indicate considerable variation over the period since counting started. Numbers dropped markedly from over 3000 fish in 1966 to less than 1000 in 1969 after which they remained relatively low. Since the power station has been operating there since 1948 these changes, if real, are probably related to some other factor and may even be due to inadequate counting, at least in some years.

Loch Shiel

Relatively little is known about fish in Loch Shiel for the loch is not heavily fished, nor are there any counting devices in the area, nor was it included in the series of lochs which were echo sounded and gill netted in 1966. Only five species are known to occur there (Table 9.2) – all of these fish with marine affinities. Surprisingly, there are no records of charr from this loch, and if this is so the community is an unusual one among the larger Scottish lochs (Campbell 1979).

The rod catch of salmon from the Shiel district (Fig. 9.7a) is a low one with a slight upward trend over the last twenty years – probably due to more anglers fishing rather than any real change in numbers. As in most other large lochs, the mean weight of fish has dropped significantly over this period (Fig. 9.7b).

Discussion

A number of features of interest has arisen from this comparative study of the fish of these large lochs. Among them, Loch Lomond (Fig. 9.10) is outstanding in the variety of its fish community (fifteen species), and indeed has more species than any other Scottish loch – all of them apparently native (Maitland 1972a). The reasons for this in the present context are probably twofold. Firstly, it is not only the largest loch by surface area but it also has a very varied character and catchment area with a wide range of habitats for different fish. Secondly, it is much further south than any of the other large lochs (Fig. 9.1) and is likely to be the first of the five lochs colonised by species new to the area, all of which are gradually dispersing northwards, and indeed have been since the end of the last ice age (Maitland 1977).

Loch Lomond is also unique among these lochs in possessing powan, known elsewhere in Scotland only from Loch Eck. This species is mentioned as occurring in Loch Morar by Murray & Pullar (1910), but there is no actual evidence of its presence there. Surprisingly, Loch Lomond and Loch Shiel

247

have no charr populations. In Loch Lomond this may be related to the enormous numbers of lampreys (Maitland 1979) or the presence of powan, for Toivonen (1972) has shown that whitefish introduced into charr lakes in Sweden can eliminate charr there – probably by direct competition for food resources (Maitland 1970). Certainly whitefish (powan) in Loch Lomond occupy the plankton feeding niche which would normally be filled by charr. The absence of the latter species from Shiel (if true) has no obvious explanation. An old report of charr in Loch Lomond must be discounted (Bidie 1896, Brown 1896).

The absence of migratory species, notably sea lamprey, river lamprey and flounder from most of the large lochs other than Loch Lomond is also rather surprising. All four of the lochs have relatively easy access rivers to and from the sea, which are certainly negotiated by salmon, trout and eels. It seems possible that the absence of sufficient suitable sandy substrate in the outflows and much of the lochs themselves might make them unsuitable for flounders. Similarly there is probably little suitable nursery substrate for the ammocoete larvae of lampreys in most of the inflows to these lochs.

The biological effects of water utilisation by hydro-electric schemes in relation to fisheries in Scotland have been reviewed by Berry (1955) and Elder (1965). Neither worker considered pumped-storage schemes in detail, and their effect on fisheries – especially those for salmonids – is uncertain. The present study indicates that the schemes at Lochs Awe and Ness have had no major impact on the fisheries there for migratory fish. However, further work is needed on the effect of these power stations on the populations of resident fish, especially charr, whose young stages are particularly vulnerable (Snyder 1975). The relationship between stocks in the lochs and their upper reservoirs also requires investigation.

Table 9.7 Summary of data (and their ranking) related to fish, especially salmon, in the large lochs. The salmon data are means for the years 1952–76 for the relevant fishery districts

Feature	Lomond	Awe	Ness	Morar	Shiel
Number of fish species	15 (1)	8 (2)	7 (3)	5 (4)	5 (4)
Number of salmon caught	* (3)	* (2)	* (1)	* (5)	* (4)
Salmon per km² catchment	* (2)	* (1)	* (3)	* (5)	* (4)
Mean weight of early salmon (kg)	5.07 (2)	5.05 (3)	5.68 (1)	4.08 (5)	4.91 (4)
Mean weight of late salmon (kg)	3.56 (4)	3.74 (3)	4.13 (1)	3.04 (5)	3.80 (5)
Mean weight of all salmon (kg)	4.10 (2)	3.95 (3)	4.64 (1)	3.26 (5)	3.93 (4)
Precedence totals	(14)	(14)	(10)	(29)	(22)

It has been possible in other papers in this series (e.g. Maitland *et al.* 1981) to rank these large lochs in respect of a number of features of their physics, chemistry and biology. Such comparisons are more difficult with this study where few really comparable data for fish are available from all the lochs. Some comparisons are shown in Table 9.7 where the lochs are ranked according to their precedence in terms of numbers of species of fish, etc. The results show that Lomond, Awe and Ness are for the most part much richer and more productive than Morar and Shiel. It must be remembered that some of the features compared here (e.g. numbers of salmon) are probably related more to the whole catchments rather than the lochs themselves, and that Morar and Shiel have much smaller catchments than the other three lochs.

Acknowledgements

This study is part of a research programme (ITE Projects 545 and 546) carried out under contract to the Nature Conservancy Council and the North of Scotland Hydro-Electric Board. We are grateful to Mr R. N. Campbell of NCC and Mr D. Simpson of NSHEB for supplying certain data. Valuable information was also contributed by the Department of Agriculture and Fisheries for Scotland (Dr K. Mitchell and Mr R. B. Williamson) the Loch Lomond Angling Improvement Association (Mr D. MacDonald), Ford Hotel (Mr D. Murray), Portsonachan Hotel (Mr R. Buchanan), the former Lochawe Hotel (Mr J. Kennedy), and Brig. R. W. L. Fellowes (Cladich fishery). Dr H. D. Slack, Mr J. D. Hamilton and the late Mr F. W. K. Gervers collected much of the Loch Lomond data from 1950–60, and Mr D. N. Webster helped enormously with the field work at Craigroyston during 1978–79

References

Anonymous, 1972. Loch Morar Survey. Ann. Rep. 1971, 1–12.
Bailey-Watts, A. E. & Duncan, P., 1981a. The ecology of Scotland's largest lochs: Lomond, Awe, Ness, Morar and Shiel. Ed. P. S. Maitland. Chemical characterisation – a one year comparative study. Chap. 3. This volume.
Bailey-Watts, A. E. & Duncan, P., 1981b. The ecology of Scotland's largest lochs: Lomond, Awe, Ness, Morar and Shiel. Ed. P. S. Maitland. The phytoplankton. Chap. 4. This volume.
Bailey-Watts, A. E. & Duncan, P., 1981c. The ecology of Scotland's largest lochs: Lomond, Awe, Ness, Morar and Shiel. Ed. P. S. Maitland. A review of macrophyte studies. Chap. 5. This volume.
Berry, J., 1955. Hydro-electric development and nature conservation in Scotland. Proc. R. phil. Soc. Glasg. 77: 23–36.

Bidie, G., 1896. Char in Loch Lomond. Ann. Scott. Nat. Hist. 124: 258.

Brown, A., 1891. The fishes of Loch Lomond and its tributaries. Scot. Nat. 10: 114–124.

Brown, A., 1896. Char in Loch Lomond. Ann. Scott. Nat. Hist. 123: 192.

Calderwood, W. L., 1921. The salmon rivers and lochs of Scotland. London: Arnold.

Campbell, R. N., 1979. Ferox trout, *Salmo trutta* L., and charr, *Salvelinus alpinus* (L.) in Scottish lochs. J. Fish. Biol. 14: 1–29.

Castle, P., 1937. Angling holidays in Scotland. Edinburgh: Oliver & Boyd.

Copland, W. O., 1956. Notes on the food and parasites of Pike *(Esox lucius)* in Loch Lomond. Glasg. Nat., 17: 230–235.

Elder, H. Y., 1965. Biological effects of water utilisation by hydro-electric schemes in relation to fisheries, with special reference to Scotland. Proc. R. Soc. Edinb., B, 69: 246–271.

Fuller, J. D., Scott, D. B. C. & Fraser, R., 1976. The reproductive cycle of *Coregonus lavaretus* (L) in Loch Lomond, Scotland, in relation to seasonal changes in plasma cortisol concentrations. J. Fish. Biol., 9: 105–117.

Gervers, F. W. K., 1954. A supernumerary fin in the powan *(Coregonus clupeoides* Lacepede). Nature, Lond., 174: 935.

Hunter, W. R., Slack, H. D. & Hunter, M. R., 1959. The lower vertebrates of the Loch Lomond district. Glasg. Nat. 18: 84–90.

Lamond, H., 1922. Some notes on two of the fishes of Loch Lomond: the Powan and the Lamprey. Rep. Fishery Bd., Scot. 2: 1–10.

Lamond, H., 1931. Loch Lomond. Glasgow: Jackson.

Lumsden, J. & Brown, A., 1895. A guide to the natural history of Loch Lomond and neighbourhood. Glasgow: Bryce.

MacDonald, T. H., 1959. Estimates of length of larval life in three species of lamprey found in Britain. J. Anim. Ecol. 28: 293–298.

McNiven, A., 1863. Note on the powan. Proc. Nat. Hist. Soc. Glasg., 1: 48–49.

Maitland, P. S., 1965. The feeding relationships of salmon, trout, minnows, stone loach and three-spined sticklebacks in the River Endrick, Scotland. J. Anim. Ecol. 34: 109–133.

Maitland, P. S., 1967a. The artificial fertilisation and rearing of the eggs of *Coregonus clupeoides* Lacepede. Proc. R. Soc. Edinb. 70: 82–106.

Maitland, P. S. 1967b. Echo sounding observations on the Lochmaben vendace, *Coregonus vandesius* Richardson. Trans. J. Dumfries Galloway nat. Hist. Antiq. Soc. 44: 29–46.

Maitland, P. S., 1969. The reproduction and fecundity of the powan, *Coregonus clupeoides* Lacepede, in Loch Lomond, Scotland. Proc. R. Soc. Edinb., B, 70: 233–264.

Maitland, P. S., 1970. The origin and present distribution of *Coregonus* in the British Isles. Proc. Int. Symp. Biol. Coregonid Fish 1: 99–114.

Maitland, P. S., 1972a. Loch Lomond: man's effects on the salmonid community. J. Fish. Res. Bd. Can. 29: 849–860.

Maitland, P. S., 1972b. A key to the freshwater fishes of the British Isles with notes on their distribution and ecology. Freshwater Biological Association, Scientific Publication, No. 27.

Maitland, P. S., 1976. Fish in the large freshwater lochs of Scotland. Scott. Wildl. 12: 13–17.

Maitland, P. S., 1977. Freshwater fish in Scotland in the 18th, 19th and 20th centuries, Biol. Conserv. 12: 265–278.

Maitland, P. S., 1979. Scarring of whitefish *(Coregonus lavaratus)* by lampreys *(Lampetra fluviatilis)* in Loch Lomond. Can. J. Fish. Aquat. Sci. 37.

Maitland, P. S., 1981. The ecology of Scotland's largest lochs: Lomond, Awe, Ness, Morar and Shiel. Ed. P. S. Maitland. Introduction and catchment analysis. Chap. 1. This volume.

250

Maitland, P. S., Smith, B. D. & Dennis, G. M., 1981. The ecology of Scotland's largest lochs: Lomond, Awe, Ness, Morar and Shiel. Ed. P. S. Maitland. The crustacean zooplankton. Chap. 6. This volume.

Mills, D. H., 1969. The growth and population densities of roach in some Scottish waters. Proc. Brit. Coarse Fish Conf. 4: 50–57.

Marcy, B. C., 1973. Vulnerability and survival of young Connecticut River fish entrained at a nuclear power plant. J. Fish. Res. Bd. Can. 30: 1195–1203.

Murray, J. & Pullar, L., 1910 Bathymetrical survey of the freshwater lochs of Scotland. Edinburgh: Challenger. Vols 1–6.

Nall, G. H., 1933. Sea trout of the River Leven and Loch Lomond. Rep. Fishery Bd. Scot. 4: 1–22.

North of Scotland Hydro-Electric Board, 1979. Power from the glens. Edinburgh: NSHEB.

Parnell, R., 1838. Observations on the Coregoni of Loch Lomond. Ann. Nat. Hist. (Lond.), 1: 161–165.

Robertson, D., 1870. On *Petromyzon fluviatilis*, and its mode of preying on *Coregonus clupeoides*. Proc. Nat. Hist. Soc. Glasg. 2: 61–62.

Robertson, D., 1888. The pike, *Esox lucius*, Lin. 1888. Trans. Nat. Hist. Soc. Glasg. 2: 212–214.

Scott, D. B. C., 1975. A hermaphrodite specimen of *Coregonus lavaretus* (L.) (Salmoniformes, Salmonidae) from Loch Lomond, Scotland. J. Fish Biol. 7: 709.

Shafi, M., 1969. Comparative studies of populations of perch (*Perca fluviatilis* L.) and pike (*Esox lucius* L.) in two Scottish lochs. Ph.D. Thesis, University of Glasgow.

Shafi, M. & Maitland, P. S., 1971a. The age and growth of perch, *Perca fluviatilis* L. in two Scottish lochs. J. Fish. Biol. 3: 39–57.

Shafi, M. & Maitland, P. S., 1971b. Comparative aspects of the biology of pike, *Esox lucius* L., in two Scottish lochs. Proc. R. Soc. Edinb. B, 71: 41–60.

Slack, H. D., 1955. Factors affecting the productivity of *Coregonus clupeoides* Lacepede in Loch Lomond. Verh. int. Verein. theor. angew. Limnol. 12: 183–186.

Slack, H. D., 1957. Studies on Loch Lomond. I. Glasgow: Blackie & Son.

Smith, B. D., Maitland, P. S., Young, M. R. & Carr, M. J., 1981a. The ecology of Scotland's largest lochs: Lomond, Awe, Ness, Morar and Shiel. Ed. P. S. Maitland. The littoral zoobenthos. Chap. 7. This volume.

Smith, I. R. & Lyle, A. A., 1979. The extent and distribution of fresh waters in Great Britain. Cambridge: Institute of Terrestrial Ecology.

Smith, I. R., Lyle, A. A. & Rosie, A. J., 1981b. The ecology of Scotland's largest lochs: Lomond, Awe, Ness, Morar and Shiel. Ed. P. S. Maitland. Comparative physical limnology. Chap. 2. This volume.

Snyder, D. E., 1975. Passage of fish eggs and young through a pumped storage generating station. J. Fish. Res. Bd. Can. 32: 1259–1266.

Stirling, J., 1931. Fishing for trout and seatrout with worm and wet fly. Glasgow: Allan.

Toivonen, J., 1972. The fish fauna and limnology of large oligotrophic glacial lakes in Europe (about 1800 AD). J. Fish. Res. Bd. Can. 29: 629–637.

Wood, I., 1954. Loch Lomond and its salmon. Glasgow: Scottish Field.

Young, J., 1870. Note on Tench in Loch Lomond. Proc. Nat. Hist. Soc. Glasg. 2: 67.

10.   Comparisons and synthesis

P. S. Maitland, I. R. Smith, A. E. Bailey-Watts,
B. D. Smith & A. A. Lyle

Abstract

Available data on the ecology of these five large lochs are reviewed. Climatic factors have less influence than morphometric differences among the lochs, where mean depth and shoreline slope are among the most important structural features. The chemical characteristics of the lochs are clearly determined by the nature of the geology, land use and population densities of the catchments. The physico-chemical characteristics influencing biological diversity and production tend to favour some lochs (e.g. Lomond) more than others (e.g. Morar) and though all five lochs would conventionally be classified as oligotrophic, the cumulative effects of relatively small differences tend to make biological distinctions greater than might be otherwise expected. The unique dual character of Loch Lomond is emphasised and its north and south basins are contrasted. The status of the five lochs as unstressed ecosystems compares favourably with relevant waters elsewhere in the world. From both resource and conservation viewpoints all five lochs are of importance and their future status should be conserved by developing integrated multi-purpose plans for each loch and its catchment.

Introduction

The nine previous studies in this series (e.g. Maitland 1981, Smith *et al.* 1981b, etc.) have each been concerned with a different facet of the ecology of the five lochs concerned – Lomond, Awe, Ness, Morar and Shiel. The papers have dealt mainly with between-loch comparisons and speculated on possible reasons for some of the differences which are apparent among these large lochs. The intention of the present paper is to integrate and discuss some of the main features of the studies and to examine further comparative aspects of the physico-chemical and biological features of Scotland's five largest lochs. In addition, a few comparative international observations are made giving some perspective on these lochs in a world context. Their limnological classification is then examined on the basis of published criteria, and finally the status and conservation of the lochs from the viewpoint of their importance as a national resource is discussed.

Only five of the large lochs of Scotland are included in this comparative study (Fig. 10.1), but there are many others of comparable size in various parts of the country (mainly the highlands). In fact, in a consideration of the five size parameters of surface area, length, volume, maximum depth and mean depth, a total of 29 lochs are involved each of which is bigger in at least

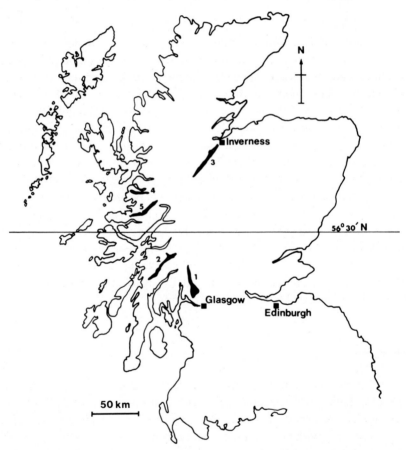

Fig. 10.1 Outline diagram of Scotland showing the location of Lochs (1) Lomond, (2) Awe, (3) Ness, (4) Morar, and (5) Shiel.

254

Table 10.1 The largest Scottish lochs: morphometric data on the 29 largest lochs in Scotland by area, volume, length and depth (maximum and mean). The data are from Murray & Pullar (1910)

	Loch	Area (km²)	Loch	Volume (m³ . 10⁶)	Loch	Length (km)	Loch	Max. depth (m)	Loch	Mean depth (m)
1.	Lomond	71.1	Ness	7452.0	Awe	41.0	Morar	310.3	Ness	150.7
2.	Ness	56.4	Lomond	2628.0	Ness	39.0	Ness	229.8	Morar	86.6
3.	Awe	38.5	Morar	2307.3	Lomond	36.4	Lomond	189.9	Lochy	69.8
4.	Maree	28.6	Tay	1601.3	Shiel	28.0	Lochy	161.8	Treig	63.2
5.	Morar	26.7	Awe	1230.4	Shin	27.7	Ericht	156.0	Katrine	60.7
6.	Tay	26.4	Maree	1091.3	Tay	23.4	Tay	154.8	Tay	60.7
7.	Shin	22.5	Ericht	1076.8	Ericht	23.3	Katrine	150.9	Ericht	57.7
8.	Shiel	19.6	Lochy	1068.3	Maree	21.7	Rannoch	134.1	Rannoch	51.0
9.	Rannoch	19.1	Rannoch	973.7	Arkaig	19.3	Treig	132.9	Glass	48.4
10.	Ericht	18.7	Shiel	792.5	Morar	18.8	Shiel	128.0	Arkaig	46.5
11.	Arkaig	16.2	Katrine	772.3	Lochy	15.7	Maree	111.9	Earn	42.0
12.	Lochy	15.3	Arkaig	752.5	Rannoch	15.6	Glass	111.3	Shiel	40.5
13.	Leven	13.7	Earn	408.4	Katrine	12.9	Arkaig	109.4	More	38.4
14.	Katrine	12.4	Treig	393.8	Langavat	12.6	More	96.3	Maree	38.2
15.	Earn	10.1	Shin	350.6	Laggan	11.3	Awe	93.6	Morie	38.2
16.	Harray	9.8	Fannich	309.2	Quoich	11.2	Earn	87.5	Lomond	37.0
17.	Fannich	9.3	Assynt	247.2	Fannich	11.1	Assynt	86.0	Muick	35.5
18.	Fionn	9.1	Quoich	236.3	Earn	10.4	Fannich	86.0	Fannich	33.2
19.	Langavat	8.9	Glass	234.6	Assynt	10.2	Quoich	85.7	Suainaval	33.1
20.	Assynt	8.0	Fionn	160.5	Naver	9.9	Morie	82.3	Awe	32.0

255

one of these measurements than one of the five studied here (Table 10.1). However, in overall size Loch Ness must probably be considered as the largest of Scottish lochs, its magnitude within Great Britain being emphasised by the fact that it holds more water than all the lakes and reservoirs in England and Wales put together (Smith & Lyle 1979).

Diversity and abundance

The overall diversity of the flora and fauna of the five lochs is a matter of some interest and a summary of some of the data concerned is given in Fig. 10.2. Possible reasons for some of the apparent differences are discussed below. In general the situation with diversity is similar to that for overall richness, with Lochs Lomond and Awe possessing the widest range of plants and animals and Lochs Morar and Shiel the lowest. Loch Ness is somewhat intermediate between these two pairs. For some groups of organisms (e.g. crustacean zooplankton) there is relatively little difference in the species and communities found in the lochs. In others (e.g. fish), the communities of some lochs (Loch Lomond) are considerably more diverse than others (Loch Shiel).

The reasons for the differences among the lochs may vary from group to group of plants or animals. Many species of zooplankton are well known to be cosmopolitan. With benthic invertebrates, diversity for any loch as a whole may well be related to the diversity of the available habitat and this is probably reflected in the fact that Loch Lomond, which is easily the most diverse of the large lochs from the point of view of physical habitats has also a fairly diverse benthic flora and fauna.

A number of different measures of the standing crops and other indices of richness of the lochs has been obtained during the research programme. Some of the most relevant measurements ranging through the trophic levels of the ecosystems are presented in Fig. 10.2. There are some anomalies, but there is no doubt that in most respects Lochs Lomond and Awe are the richest and Lochs Morar and Shiel the poorest, with Loch Ness somewhat intermediate. There is also a reasonable relationship between the physical and chemical parameters pertaining to richness and the amounts of plants and animals found.

Although on a world scale the five lochs appear to be similar (see below), in the Scottish or British context they have some important distinguishing features. These are accounted for by the differing physico-chemical conditions in the lochs. In broad terms the differences are not great. However, there is a tendency for those characteristics that favour biological productivity

256

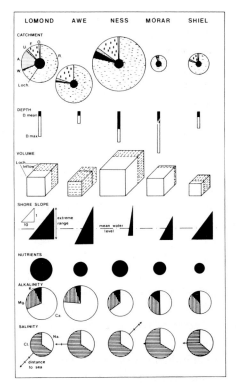
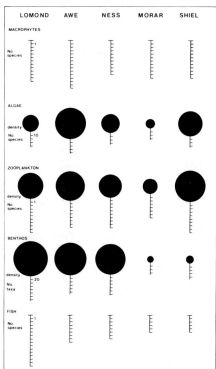

Fig. 10.2a (left)   Relative values of some physical and chemical features of the five lochs.

Catchment: Areas of the circles are proportional to the loch and catchment areas. Land use characteristics of the catchment are indicated. (R) rough grazing. (W) standing water. (A) arable land. (U) urban. (F) forest. (O)other.

Depth: Mean and maximum depths are indicated.

Volume: The volume contained by the solid lines indicates the loch volume, the hatched volume indicating the mean annual inflow.

Shore slope: The vertical face of the triangle indicates the extreme range of water level while the horizontal edge represents the width of affected shore.

Nutrients, alkalinity and *salinity:* The areas of the circles are proportional to mean concentrations in the loch water. Nutrients refers to total nitrogen only. Note that the scales for the various elements are not the same, the ratios between them being 0.001:2:1 respectively.

The distance to sea scale also indicates the direction of the shortest distance to the open sea.

Fig. 10.2b (right)   Relative values of some biological features of the five lochs.

The density units for the biological features are as follows:

Algae individuals/l

Zooplankton numbers/unit volume

Benthos numbers/10 min shore sample

Note that the vertical scale of species numbers varies.

257

to be more pronounced in some lochs than others. For example Lochs Lomond and Awe have the lowest mean depths and the highest proportions of base rich geology and arable land in their catchments. These two lochs have other limnological features in common: e.g. reduced mean thermocline depths, greater water level and temperature variability and large inflows. The greatest variety of catchment types occurs around Loch Lomond, followed by Lochs Awe and Ness. There are similarly extensive fringes of deciduous trees around Lochs Lomond and Awe (71% and 72% of the shore line respectively).

Water chemistry in these two lochs is also similar. As a result of the present study the lochs were divided into alkaline and saline types. Lochs Lomond and Awe are both of the former group, with the highest alkalinities and pH values in the series and the highest calcium and magnesium values (Bailey-Watts & Duncan 1981). Considering all these features together it is not too surprising that Lochs Lomond and Awe are the richest lochs and have many similarities between their faunas and floras.

In attempting to identify the causes for the observed differences between the lakes in water chemistry and phytoplankton, indices of potential loch trophic status have been used. These indices are based on features of the catchments: (1) total area, (2) the area devoted to arable farming, (3) the area occupied by base rich rock and (4) the human population density in the catchment. To examine the relative potential influences of these features on each loch they are related to the loch total volume and the epilimnion volume.

Fig. 10.3 shows bar charts of the various ratios (e.g. catchment area to loch

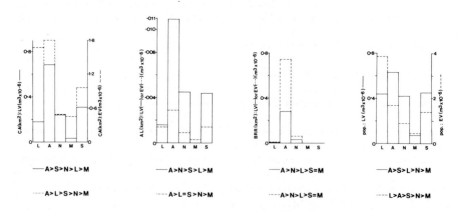

Fig. 10.3 Bar charts of various ratios relevant to the potential nutrient richness of Lochs Lomond (L), Awe (A), Ness (N), Morar (M) and Shiel (S), and the ranking of the lochs according to these ratios. Area of catchment (CA), its arable land (AL) and its base-rich rock (BRR) and the human population (pop.) are each related to total loch volume (LV) and the volume of the epilimnion (EV) at the height of summer stratification.

258

total and epilimnion volumes) and the ordering of the lochs starting with those exhibiting the highest values. The ratios of the four measures related to total loch volume consistently indicate that Loch Awe is potentially the richest and Loch Morar the poorest, although it shares this bottom position with Loch Shiel as the catchment areas of both lochs appear to be virtually devoid of arable land. Whilst the upper and lower limits of the ranges found can be established the form of the relationship described by intermediate values is less clear. Thus, the lochs occupying the 2nd, 3rd or 4th places in the rankings vary with the index used. A very notable feature is that the north basin of Loch Lomond which on the basis of numerous studies is usually considered the richest (or second richest) loch – appears to have greatest affinities with the 'poor' Lochs Shiel and Morar (depending on the measure of potential richness used). Thus its water quality (chemical and biological) does not appear to reflect what would be expected from its catchment characteristics (see below).

The other remarkable observation is that the ratios obtained for Loch Shiel merit its position in the centre or towards the rich end of the rankings. This is not in keeping with its water chemistry but does relate better to its relative richness based on phytoplankton abundance. The very deep and humic-stained Loch Ness is usually at least midway in, if not towards the rich end of, the orderings and this is also in keeping with the status of its phytoplankton, which is richer than might have been expected from data on the loch basin alone.

In order to distinguish the lochs on the basis of phytoplankton abundance the data relating to the summer stratification periods have been used. Outwith these periods the distinctions are less marked. From the summer algal development Loch Awe appears the richest in that the crop maximum achieved is greatest of all the lochs, and moreover the increase to this maximum commences much earlier than in the other lochs. In order of decreasing algal abundance Loch Awe is followed by Lochs Lomond, Ness, Shiel and Morar. The 'elevated' position of Lomond is reflected in at least some of the rankings shown in Fig. 10.3 where the indices of potential richness are related to epilimnion volume.

The diversity and abundance of benthic plants and animals can be greatly affected by substrate, this in some cases overriding other 'large scale' physico-chemical conditions. For example, despite the similarity of the water chemistry and catchments of Lochs Morar and Shiel and their close proximity, Loch Shiel has more species of littoral zoobenthos in common with Loch Lomond than with Loch Morar. Loch Shiel is akin to Loch Lomond (and Loch Awe) with respect to depth and shore slope and the similarity of its

259

substrate to that of Loch Lomond (but not of Loch Awe where stones and boulders form a large part of the shore) has been pointed out by Smith *et al.* (1981a).

This predominance of stony substrates in Lochs Awe, Ness and Morar (around 80% of the littoral zone) can also affect benthos and macrophytes. On these stony wave-washed shores little colonisation by larger plants can take place. Where aquatic plants do grow, in sheltered bays, then very high numbers of zoobenthos occur. Many species found in these bays are atypical of the loch as a whole and are more characteristic of mesotrophic conditions. Thus at Inchnacardoch Bay (Loch Ness) large numbers of *Asellus* spp. were collected. In the sheltered bay at the south end of Loch Awe larvae of the weed dwelling mayfly *Cloeon* reached densities of more than 12.000 per 10 min collection. Loch Lomond has the smallest relative area of stony shore. The south basin in particular has extensive shores whose substrates have smaller particle sizes and comparatively large areas colonised by higher aquatic plants. It also has the greatest abundance and diversity of littoral zoobenthos.

When substrates are similar (e.g. the greater parts of Lochs Morar and Ness) then other physico-chemical characteristics can affect the fauna. The great depths, steep shores and wave heights of Lochs Ness and Morar are somewhat alike. These physical similarities are reflected in the benthic faunas with typically lotic species being regularly collected in both lochs. The steep wave-washed shores approximate to stony streams and the occurrence of a species normally associated with high altitude streams, *Ameletus inopinatus* Eaton, is characteristic of Lochs Ness and Morar. The enormous volumes and northern latitutdes of these lochs also means that summer water temperatures are never high. The maximum surface water temperatures recorded in this study were 14.0 °C in Loch Ness and 14.7 °C in Loch Morar.

With the fish fauna, on the other hand, the history of colonisation of each loch following the last glacial period is probably the most important single factor affecting diversity at the present time. Thus the main reason for the high number of fish species associated with Loch Lomond, is the fact that this loch is easily the most southerly of the five (Fig. 10.1) and is within, but at the northern edge of, the present distribution range of several fish species (Maitland 1972a). Roach, *Rutilus rutilus* (L.) and stone loach, *Noemacheilus barbatulus* (L.) are good examples of these species: for both, Loch Lomond is the most northerly population on the west side of Scotland. These and several other species seem to be gradually moving north in Great Britain and the fish communities of all the lochs are likely to become more diverse in time.

260

Comparisons with other waters

In European terms the five lochs considered in this study must still rank as relatively large, at least in some respects. Loch Morar, for instance, is the seventh deepest lake in Europe. Within the world as a whole, however, the Scottish lochs cannot really be considered as large. Although four of them (Lochs Lomond, Awe, Ness and Morar) rank within the largest 150 (as far as surface area is concerned), the area of even the largest (Lomond: 71 km^2) is less than one hundredth of at least 20 of the world's gigantic standing waters. The largest of these is Lake Superior (83 300 km^2).

The areas and mean depths of the five lochs are compared with some of the major lakes of the world in Fig. 10.4. In terms of area, the Scottish lochs are three or four orders of magnitude less than the truly large lakes and the use of large outside the Scottish context hardly seems warranted. In terms of depth, however, the differences are much less and it is the depth of the Scottish lochs that not only justifies describing them as large but which also dominates their limnological and biological features.

The zoobenthos of the five lochs is similar if examined in a world context and only the southern basin of Loch Lomond would not be classified as oligotrophic (see below). For the five lochs in general insects form the greater part of the total fauna. This dominance of the fauna by insects is typical of unproductive water bodies. With increasing productivity triclads, leeches

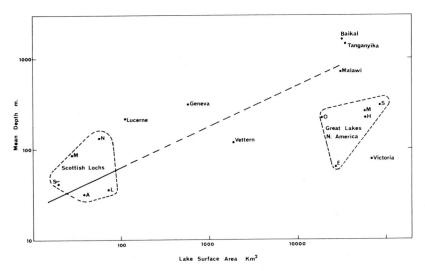

Fig. 10.4 Comparison of area and mean depth for a number of lakes throughout the world. The line is the regression of mean depth on area for rock basins in Scotland derived by Gorham (1958).

Table 10.2 Numbers of species of Tricladida, Hirundinea and Gastropoda in various lakes. Data, other than from the large lochs, are from Macan (1970) and Rigler (1972)

Lake	Country	Tricladida	Hirundinea	Gastropoda
Char	Canada	0	0	0
Shiel	Scotland	1	0	4
Morar	Scotland	1	1	1
Wastwater	England	1	1	2
Awe	Scotland	1	6	4
Ness	Scotland	5	2	2
Lomond	Scotland	4	2	7
Windermere	England	5	8	10
Esrom	Denmark	7	8	19

and molluscs form major components of the fauna. Table 10.2 compares the species of triclads, leeches and gastropods found in the five lochs with some other British, European and North American lakes.

At the extreme oligotrophic end of the scale is Char Lake in Canada, a high Arctic water body. No species of snail, flatworm or leech were recorded from this lake during an intensive survey (Rigler 1972). Loch Morar is Britain's most extreme oligotrophic loch (Ratcliffe 1977). Its unproductive status is reflected in the presence of only a single species of each of three groups (triclads, leeches and molluscs) characteristic of productive eutrophic waters. This poverty is even more striking when the abundance of these animals is considered. Leeches and triclads are each represented by only one specimen of a single species during a twelve month sampling programme. The mollusc *Lymnaea peregra* (Muller) was collected relatively commonly during the monthly sampling, however its maximum number reached only 9/10 min collection. This species is the commonest and most abundant gastropod collected in British fresh waters (Macan 1977). Loch Shiel and Wastwater are similarly unproductive lakes. The latter was classified by Pearsall (1921) as one of the most unproductive lakes in the English Lake District. Like Loch Morar, Wastwater was classified as a Grade 1 site for conservation (Ratcliffe 1977) while Loch Shiel is a Grade 2 alternative for Loch Morar.

The littoral zoobenthos of Loch Ness is more abundant than that of these extreme oligotrophic sites. However, triclads, leeches and gastropods are infrequent and the fauna is again dominated by insects. Exceptionally large numbers of triclads were collected at a site which was enriched with sewage effluent. This site was notable for another reason for the triclad found, *Phagocata woodworthi* Hymen, is a North American species new to Britain (Reynoldson *et al.* 1981).

Loch Lomond is the most productive of the five Scottish lochs, with the greatest variety and abundance of species of zoobenthos. Its south basin may be described with some justification as mesotrophic. In many respects it is similar to Lake Windermere in the English Lake District. However both these lakes are considerably less eutrophic than Lake Esrom in Denmark which, for example, has at least 19 species of snail (Table 10.2).

Although the fish fauna of these large lochs seems impoverished in comparison with large tropical lakes, all of them contain more species than the extreme northern lakes of the Holarctic region (e.g. Char Lake in northern Canada with only one species). The diversity is comparable to that found in other large British lakes – e.g. Loch Leven (8 species), Windermere (9) – though rather less than is recorded for the large lakes of central Europe: e.g. Lago Maggiore (19) and Lac Leman (23). Some of the large lakes of north-western Canada – e.g. Keller Lake (13), Babine Lake (14) – have comparable fish species diversities to Loch Lomond (15).

Lake classification

Much of the discussion above and in previous papers has centred on comparing and contrasting the five lochs. However, it is clear that in broad terms they are relatively very similar, and indeed in this context it is of interest to see how they fit into conventional schemes of lake classification. Standing waters as a whole, of course, exhibit a great variety of types, ranging from small shallow temporary pools, through ponds and lakes to enormous bodies of water which may be over 20 000 km² in surface area and over 500 m in maximum depth. Superimposed on the size and shape of the water basin are important regional differences, especially those related to geochemistry and climate. In their individuality, lakes may be compared to oceanic islands: just as an island presents many peculiarities in its rocks, soil fauna and flora due to its isolation by the ocean, so do many lakes present individuality and special peculiarities in their physical, chemical and biological features owing to their position relative to drainage from the surrounding land, and their separation from other waters by the land mass.

Some of the best known schemes of lake classification are shown in Table 10.3, and the position in them of each of the five large lochs is indicated. Usually, all five lochs always fall into exactly the same class, which may be generally described as the temperate oligotrophic type. The few exceptions are in the classification by origin (Loch Ness has a different geological origin from the other four lochs) and in the biological classification of the trophic

Table 10.3 The position of Lochs Lomond, Awe, Ness, Morar and Shiel within various systems of lake classification

Type & author	System		Position of lochs
Origin	Rock	Solution	—
		Volcanic	—
Murray & Pullar (1910)		Glacial	Lomond, Awe, Morar, Shiel
		Tectonic	Ness
		Meteor	—
	Barrier		—
	Organic		—
Thermal	Amictic (polar)		—
Hutchinson (1957)	Cold monomictic (arctic)		—
	Dimictic (temperate)		Lomond, Awe, Ness, Morar, Shiel
	Warm monomictic (tropical)		—
	Oligomictic (equatorial)		—
Hardness	Nutrient-poor (< 15 mg l^{-1} $CaCO_3$)		Lomond, Awe, Ness, Morar, Shiel
Spence (1967)	Intermediate (15-60 mg l^{-1} $CaCO_3$)		—
	Nutrient-rich (> 60 mg l^{-1} $CaCO_3$)		—
Nutrient loading	Oligotrophic		Lomond, Awe, Ness, Morar, Shiel
Vollenweider (1968)	Mesotrophic		—
	Eutrophic		—
Phytoplankton	Oligotrophic		Lomond, Awe, Ness, Morar, Shiel
Sakamoto (1966)	Mesotrophic		—
	Eutrophic		—
Invertebrates	Oligotrophic		Awe, Ness, Morar, Shiel
Thienemann (1925)	Mesotrophic		Lomond
	Eutrophic		—
	Dystrophic		—
Fish	Oligotrophic		Awe, Ness, Morar, Shiel
Maitland (1978)	Mesotrophic		Lomond
	Eutrophic		—
	Dystrophic		—

status, where there is some indication that Loch Lomond (or rather the south basin of Loch Lomond) is mesotrophic rather than oligotrophic. This fits in well with the studies discussed elsewhere where it is apparent that in the nutrient range of the five lochs concerned, Loch Lomond lies at the rich end of the spectrum.

In connection with nutrient loading, the centre of gravity of the cluster of Scottish lochs on Fig. 10.4 could be represented by a loch with a surface area of 50 km^2 and a mean depth of 70 m. Assuming phosphorus to be the limiting nutrient, the loading required to achieve mesotrophic status on the basis of the original classification by Vollenweider (1968) is 0.32 gm m^{-2} yr^{-1}. If all the

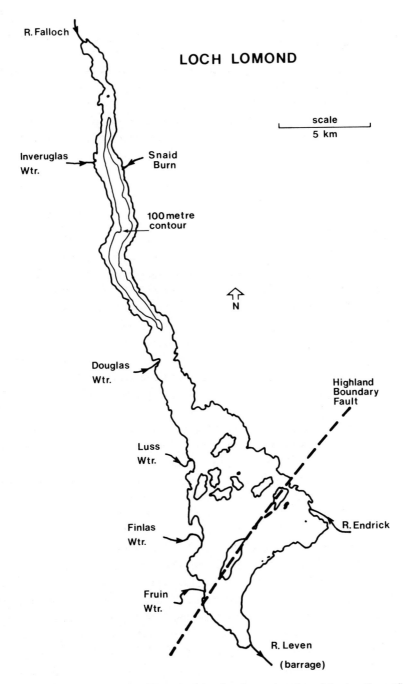

R. Falloch

LOCH LOMOND

scale
5 km

Inveruglas
Wtr.

Snaid
Burn

100 metre
contour

↑
N

Douglas
Wtr.

Highland
Boundary
Fault

Luss
Wtr.

R. Endrick

Finlas
Wtr.

Fruin
Wtr.

R. Leven
(barrage)

Fig. 10.5   Outline diagram of Loch Lomond showing the north and south basins, the outflow (River Leven) and the major inflows. Ground to the north of the Highland Boundary Fault Line is base-poor, whereas most ground to the south is base-rich.

phosphorus was derived from domestic sewage with an input of 1.5 gm per person per day, such a loch could receive the sewage from a population of 29 224 without losing its oligotrophic status. This population would have to be almost doubled for the loch to be eutrophic. The actual populations range from 12 218 within the Loch Lomond catchment to 162 in the case of Loch Morar. With the possible exception of the south basin of Loch Lomond, therefore, the depth alone virtually ensures their biological oligotrophic status.

The dual character of Loch Lomond

Throughout this series of studies Loch Lomond has shown itself to be of especial interest because of its dual character, with contrasting north and south basins (Fig. 10.5). The particular limnological relevance of this was pointed out by Slack (1957) and Russell-Hunter (1958, 1970) and is emphasised by the fact that, not only are the two basins very different physically, but the major geology and land uses within their catchments are also very different. These basic differences are reflected in various ways in the ecology of the loch and have given rise to problems in discussing it as a single unit, for example in comparing it with the other four large lochs and in classifying it in conventional limnological ways. It is pertinent here to discuss this problem further.

Many of the parameters measured during the course of these studies have been used to illustrate similarities or differences between the five lochs. As a natural consequence it has been possible also to order them in various ways. Whilst this ordering has differed slightly according to the parameter used, where the criteria indicate potential or actual richness, much the same series has resulted. Thus Lochs Awe and Lomond have been invariably identified as the rich pair of waters, Lochs Morar and Shiel as the poor lochs; Loch Ness usually tends to lie between these two extremes.

However, in a number of comparisons the results differ from this generalisation and one of these concerns Loch Lomond, or rather its north basin – to which the majority of the present investigations on temperature, stratification, chemical quality, phyto- and zooplankton have been restricted. Basically the question arises as to why Lomond north basin has appeared at or near the rich end of the rankings based on observed biomass and/or diversity of biota when a number of physical and chemical features of this basin and its catchment would suggest a lower placing in the series.

Comparative physical data on the north and south basins of Loch Lomond

Table 10.4a Morphometric and land use details of the north and south basins of Loch Lomond
and their catchments

Feature	North basin	South basin	Whole loch
Area (km²)	19.01	52.09	71.10
Volume (10^8 m³)	14.897	11.382	26.279
Mean depth (m)	78.4	21.9	31.0
Maximum depth (m)	189.9	66.1	189.9
Catchment area (km²)	271	425	696
% Arable	0.85	26.75	16.67
% Base rich rock	3.0	66.5	45.0
Human population	627	11 591	12 218
Thermocline depth(m)	10	4	7
Epilimnion volume (10^6 m³)	162	122	284

Note that the physical data refer to the basins separated at the delta of the Douglas Water (Fig.
10.5) whereas the catchment divide is south of this (Maitland 1981). The south basin here
includes both the middle and south basins of Slack (1957) and Russell-Hunter (1958).

are given in Table 10.4a. The contrast within a single system is remarkable
and almost unique for any lake in Europe. The north basin is very similar to
the other four large lochs (Awe, Ness, Morar and Shiel) involved in this study
– fjord-like, with a long, narrow, deep basin and mountainous, base-poor,
rocky catchment. The south basin on the other hand is broad and shallow
(with numerous islands) and a mainly lowland, base-rich, agricultural catchment. The physical differences between the basins are due to geological
origin, the north basin being a preglacial valley among metamorphosed rocks
of Daldradian age, the south basin (largely among sedimentary rocks of
Carboniferous and Devonian age) being created partly by 'ponding' following
glacial deposition from the north basin and partly by marine incursion inter-
glacially (Russell-Hunter 1958).

Comparative chemical data for the two basins are given by Slack (1957)
with total dissolved salt values for the north basin of 22 mg l⁻¹ and for the
south basin of 28 mg l⁻¹. Relevant data are also noted by Maitland (1972b) and
more recent data from the same source (Clyde River Purification Board)
are recorded in Table 10.4b. These show distinct differences in, for example,
alkalinity, which in the north basin ranges from 6-15 mg l⁻¹ $CaCO_3$ and is
rarely more than 13, whereas in the south basin values are always more than 8
and commonly more than 13 mg l⁻¹. pH values in the north basin are rarely
greater than 7.0, whereas in the south basin they are often more than this.
Further comparative chemistry has been carried out by Maulood & Boney

267

Table 10.4b Chemical data from the north and south basins of Loch Lomond

1. Chemical data (all in mg l^{-1}, except pH) during 1977, supplied by Clyde River Purification Board

Analysis	North basin		South basin		Whole loch
	Mean	Range	Mean	Range	
pH	6.72	6.35–7.65	7.06	6.60–7.55	6.89
Suspended solids	0.39	0–2	1.17	0–3	0.78
Oxygen (% saturation)	95	87–101	96	88–114	95.5
Nitric nitrogen (as N)	0.13	0.1–0.2	0.21	0.1–0.35	0.17
Alkalinity (as CaCO₃)	11.1	6–15	12.7	8–17	11.9
Chloride (as Cl)	6.7	4–12	6.9	4–12	6.8
Total hardness	14.9	11–26	23.2	14–77	19.1

2. Nutrients and other ions

	North basin	South basin
Nutrients		
* PO₄ · P (μg at l^{-1})	0.025–0.125	0.025–0.200
* NO₃ · N (μg at l^{-1})	6–15	10–22
** NO₃ · N (μg at l^{-1})	7–14	14–29
* SiO₂ · Si (μg at l^{-1})	2–20	2–20
Other ions		
*** Chloride (mg l⁻ l^{-1})	6–14	6–14
*** Calcium (mg Ca⁺⁺ l^{-1})	7.0	10.4
*** $\dfrac{Na + K}{Ca + Mg}$ by weight	0.33	0.27
** Alkalinity (mg CaCO₃ l^{-1})	rarely > 20	often > 20
* Alkalinity (mg CaCO₃ l^{-1})	4.5–9	7–14
* pH	6.6–7.2	6.8–7.6
** pH	rarely > 7.0	often > 7.0
*** Total minerals (mg l^{-1})	22	28

*   Maulood & Boney (1980).
**   Unpublished data (Clyde River Purification Board).
*** Slack (1957).

(1980), who confirm that the waters of the north basin are similar to, but less rich than, those of the south basin.

Seasonal comparisons of the phytoplankton in the two basins have also been made by Maulood & Boney (1980). Apart from one major difference – the greater importance of *Anabaena circinalis* (Kütz) Hansg. ex Lemm., in

268

the south basin – the species composition of the phytoplankton is similar throughout the loch. Algal abundance in the north basin (0.02–0.2 μg chlorophyll a l^{-1}) is poorer than the south (0.05 – 0.36 μg chlorophyll a l^{-1}. There are two annual peaks of chlorophyll and 'total phytoplankton units', though the first peak is earlier and larger in the south basin than in the north. There is even less difference between the two basins as far as the autumn peaks are concerned and the phytoplankton crops of the two basins are very similar in many other respects.

Though no comparative data on zooplankton are available from the literature or the present study, regular sampling was carried out from June 1955 to February 1957 by Mr J. D. Hamilton (personal communication). A summary of data for 1956 is shown in Table 10.4c. Samples were collected by towing a

Table 10.4c   Densities of zooplankton (numbers per m³) in the north and south basins of Loch Lomond. The 1956 data were supplied by J. D. Hamilton (personal communication).

Value	North (1977–78)	North (1956)	South (1956)	Mean (1956)
Minimum	133	440	460	450
Maximum	4093	3400	3740	3570
Mean	1101	1654	1734	1694

Hardy Small Plankton Sampler (Glover 1953) at a depth of 5 m for 20 min. Although the densities in the south basin were on average greater than those in the north basin, the differences were small. As with phytoplankton, there were no notable differences in species composition with *Diaptomus gracilis* Sars, *Cyclops abyssorum* Sars, *Daphnia hyalina* Leydig, *Bosmina coregoni* Baird, *Bythotrephes longimanus* Leydig and *Leptodora kindti* (Focke) being the most important species (in that order) in each basin. The north basin data are comparable to those obtained during the present work (Maitland, *et al.* 1981).

During the survey of littoral zoobenthos of all the large lochs (Smith, *et al.* 1981a) it was clear that the site sampled monthly in the north basin of Loch Lomond was impoverished when compared with south basin sites. Indeed the abundance of animals at this northern site was of the order of that found in Lochs Morar and Shiel. The site is certainly exposed and this may partly account for the paucity of the fauna. However, if the mean numbers of animals/10 min collection in north and south basin sites are compared (Table 10.4d) then clearly this is not the complete explanation. Samples taken from north basin sites have totals ranging from 129 to 2308 animals/10 min collection (mean: 939). In the south basin the total numbers collected range from

269

Table 10.4d Comparative abundances (numbers per 10 minute collection) of littoral zoobenthos from various sites in the north and south basins of Loch Lomond. The site numbers are those of Smith *et al.* (1981a)

	North basin		South basin	
	Site	Numbers	Site	Numbers
	2	2308	6	893
	39	982	9	2157
	43	441	11	9156
	45	129	12	3470
	49	902	17	2487
	51	355	22	1753
	56	1749	24	1004
	70	647	26	2849
			32	2742
			36	3828
Mean number/10 min collection		939		3034
Total number of taxa		62		82

893 to 9156/10 min collection (mean: 3034). Further qualitative differences between the two basins are apparent. In the North basin 62 taxa were identified, compared with 82 taxa in the in the south. There are larger numbers of non-insects (characteristic of productive waters) in the south basin (26 taxa) compared with 16 taxa of these groups in the north. Only in the south basin are typical lentic species (e.g. *Ischnura* and *Sigara*) found.

The profundal zoobenthos of the two basins has been compared by Slack (1965) who found both qualitative and quantitative differences. Larval Chiro-nomidae and *Pisidium* each have dominant species in the different basins: of the former *Sergentia* characterises the northern basin and *Tanytarsus* the southern. *Pisidium conventus* Clessin dominates the northern basin and is replaced by other species, especially *P. lilljeborgi* Clessin in the southern. Distinctions among the Oligochaeta are less clear, though *Stylodrilus* domin-ates in the north and *Limnodrilus* in the south. Mean numbers of all benthos found by Slack (1965) showed less contrast with densities of about 1020 m^{-2} in the north basin (with a mean dry weight of 385 mg m^{-2}) and 1480 m^{-2} in the south basin (with a mean dry weight of 235 mg m^{-2}). According to Russell-Hunter (1958) the disparity may be greater than this for he quotes densities of 25-30 animals m^{-2} in the north basin and 400-500 m^{-2} in the south basin.

Virtually all of the species of fish which are known to occur in Loch

Table 10.4e The mean number and percentage frequency of occurrence of fish in 14 comparable sets of gill nettings from the north and south basins of Loch Lomond

Species	North basin		South basin	
	Number	% Frequency	Number	% Frequency
Salmon	1	14	1	14
Trout	135	100	9	43
Powan	201	100	543	86
Pike	1	14	9	72
Roach	0	0	43	58
Perch	4	29	4	43
Flounder	1	14	1	14
Totals	343	—	610	—

Lomond (Maitland 1972b) are common in the south basin. However, although almost all of them have been recorded from the north basin too, several are uncommon there and there is a very distinct difference in the composition of the fish communities of the two basins. The results of several comparable sets of nettings in the two basins are shown in Table 10.4e and give some indication of the differences as far as the larger species are concerned. Trout, powan and perch are the three common species in the north basin, and whilst these occur commonly in the south basin too, so do pike and roach which are uncommon in the north basin.

Though few quantitative data are available, comparable catches from the south basin are normally higher than those from the north. Some aspects of the biology of the fish show no differences between the two basins, for example there seems to be no difference in the incidence of lamprey scars on powan (Maitland 1980).

The physico-chemical properties of the loch basins and respective catchments can be taken as a multi-variate index of some of the major potential controls of lake productivity. In the 'one-basin' Lochs Awe, Ness, Morar and Shiel this premise is essentially borne out by the actual measures of floral and faunal abundances and/or diversity. No such simple connection has been possible for Loch Lomond however – probably because of the influence of the richer southern basin.

The magnitude and timing of the development of, for example, stratification, might suggest considerable admixture of water between the basins. This would mask the differences in phytoplankton developments that might be otherwise expected if the two basins were separated and each controlled

271

Plate 10.1a Oblique aerial view of Loch Lomond, taken from above the south basin looking north. (Photo: Aerofilms Ltd.)

Plate 10.1b   Oblique aerial view of Loch Awe, taken from above the north-west arm, looking south-west. (Photo: Aerofilms Ltd.)

Plate 10.1c Oblique aerial view of Loch Ness, taken from above the Caledonian Canal near Fort Augustus at the south-west end, looking north-east up the loch.
(Photo: Aerofilms Ltd.)

274

Plate 10.1d Oblique aerial view of Loch Morar, taken from above the sea looking due east along the loch. (Photo: Aerofilms Ltd.)

275

Plate 10.1e Oblique aerial view of Loch Shiel, taken from near the south-west end looking north-east along the loch. (Photo: Aerofilms Ltd.)

Plate 10.2 Two oblique aerial views of Loch Lomond: (a) looking south along the narrow deep north basin to the broad shallow south basin in the distance – Ben Lomond to the left, and (b) looking north along the eastern edge of the south basin to the north basin in the distance – Ben Lomond to the right. (Photos: Stanley V. Mills.)

277

purely by their respective and contrasting morphometry, nutrient loadings, depth mixing regimes etc.

The temporal extent of plankton growth is also similar between the basins: this would not be expected between such different basins if they were completely isolated. It would appear that whilst the south basin is undoubtedly the most productive area of the loch, the potential differences between it and the northern area are not realised as mixing of the water tends to deplete crops in the south end and to raise them in the north to levels greater than might otherwise be expected.

Smith *et al.* (1981b) have shown that the mean throughflow due to the runoff from the catchments in Loch Lomond is about 45 m^3 s^{-1}, and show that, for a wind speed of 10 m s^{-1}, the water transport in the direction of the wind is of the order of 1000 m^3 s^{-1}. Admittedly, the wind speed is less than this for 80% of the time with a corresponding reduction in the wind-driven transport. The very large differences in these transport rates indicate, nevertheless, that it is wrong to think of a steady movement of water towards the outflow and that a moderate transfer of water between basins will occur quite regularly.

The effect of this mixing is less marked, however, on the zoobenthos and fish which are less directly affected by water movement and nutrient mixing. They appear to be influenced more by local factors, e.g. shore slope and substrate type, and consequently exhibit a stronger contrast between north and south basins than the plankton. In conclusion, the duality of Loch Lomond, which is so evident from its basin and catchment characteristics, is seen clearly in its zoobenthos and fish, but less so in its water chemistry and plankton, apparently because of the considerable amount of wind-induced mixing which must take place between the north and south basins.

Discussion

There are basically two types of descriptive ecological research leading towards a better understanding of freshwater systems. One type investigates, sometimes over long continuous periods, one or a few systems in as much detail as possible. The other type surveys in broad terms a large number of waters – and as a consequence each water is looked at relatively superficially. Using the first approach many studies have shown how various systems function – i.e. their physical and chemical properties and biotic components have been identified and the ways in which they interact have been described and – in some cases – quantitatively assessed.

The second broadly-based 'survey' approach also provides a description of the various physical, chemical and biological elements in a wide range of systems. However, in order to achieve its aim of a broad census of conditions in the bewildering variety of lakes it is really only in the case of the more temporally stable elements that the survey techniques can reasonably provide answers. In the absence of short time-interval sampling for example the less seasonally varying physico-chemical elements are likely to be the most valuable indicators of lake type. Judicious choice of sampling period however can allow an assessment of seasonally-varying but nevertheless strictly cyclic components, e.g. invertebrate abundance. In evolving a limnological catalogue, the broader approach begins to describe the overall range in lake types, – their origins, physical structure, chemical and biological composition. Then by association analysis at least, hypotheses can be erected on the factors controlling the observed variation in the lake communities, their productivity etc. In addition – and this is where the two main research approaches link – the results of the second approach enable the findings of the intensive 'one-site' studies to be placed in broad perspective.

The present study was designed with both types of approach in mind. In studying each of the five lochs concurrently, and in a strictly comparable way, no intensive work could be done on any single loch. However, a large number of factors have been assessed overall and these have included seasonally-varying components. The general tendency in the limnological literature has been to order lakes. As discussed above the most popular scheme concerns various features associated with trophic status. Since trophic status relates most directly to the chemical – and in particular the nutrient (N and P) – conditions it is on these that the well-known system of Vollenweider (1968) is based.

Commonly however, in addition, biological features associated with or reflecting trophy are used (Thienemann 1925) and the influence of physical features e.g. water depth (Rawson 1955) and retention time (Vollenweider 1977) on the chemical and biological results of differing nutrient levels has been recognised. What should be noted here is that many of the features used to place lakes into trophic categories are seasonally-varying.

In broad terms, the five large Scottish lochs are oligotrophic. Of the large range of existing lake types – even within the British Isles – they would occupy only a small part. What the present study has revealed however, is that in addition to the kinds of detailed features unique to any single lake, there are fundamental properties by which each of these may be distinguished and thus compared with each other. This has led to a preliminary ordering of the waters from Loch Morar at the ultra-oligotrophic end of this segment of the trophic scale to Lochs Lomond and Awe at the 'richer' end. What is satisfactory is that the same or similar ordering has been indicated by data on physical and catchment features, on chemistry, and on differing biological components.

It has already been shown that the external, climatic factors are less important than the morphometric differences between the lochs. Variation in energy supply is not, therefore, important. It is suggested that the most important structural features are the mean depth (and its related influence on thermal structure) and the shoreline slope. The chemical characteristics of the five lochs are determined primarily by the geology, land use and population of the catchments. When the rankings of these physical and chemical features are compared it is clear that there is a tendency, admittedly not always consistent, for those attributes that favour biological production and diversity to be more common in some lochs compared to others.

Perhaps the main conclusion of this comparative study is that although all five lochs would be classified similarly as deep, oligotrophic waters from the standpoint of general comparative limnology, the cumulative effect of relatively small differences between them is to make the biological distinctions greater than might be otherwise expected.

The large lochs of Scotland have been the subject of a variety of research projects for well over a hundred years. Before the bathymetric survey of Murray & Pullar (1910) a number of individual papers appeared (Christison 1871, Buchan 1872, Buchanan 1885, Morrison 1886, Murray 1887, and Robertson 1888). Apart from the bathymetric survey itself, many individual papers were published on specific topics (e.g. Watson 1903, 1904a,b, Wedderburn 1904, 1907a,b, Wedderburn & Watson 1909). Most of this work has dealt with aspects of physical limnology and such studies have continued to be a feature of research at these lochs especially Loch Ness (Mortimer 1955, Thorpe 1971, 1974, Thorpe & Hall 1974, Thorpe et al. 1972, Thorpe et al. 1977). They are likely to remain sites of major importance for limnological research of various kinds.

In resource terms all five of these large lochs (and many of the other large Scottish lochs) can only become increasingly important as the demand for large amounts of clean water increases from a very wide variety of users particularly industry and recreation. The series of studies forming the present research project has demonstrated clearly the conservation and limnological interest of all five of these lochs and the conflict of user interests – already very obvious at Loch Lomond – is likely to grow.

The high amenity value of all five of the lochs – to tourists and for recreation purposes, as fisheries and sources of water supply – means that changes detrimental to water quality are likely to be opposed. Water quality apart, however, those with interests in tourism, camping, boat hire, cabin cruising, power boating, water skiing, canoeing, bathing, natural history, angling, research, conservation and water supply are likely to be involved in a conflict of interests. The problems concerned are likely to be resolved successfully only if some form of co-ordinated programme involving the integration of multi-purpose usage is developed for each loch and its catchment. Although the situation at the less stressed lochs (Morar and Shiel) is not yet urgent, this is not the case at Loch Lomond where considerable ecological changes could take place by the end of the present century unless there is careful management.

Acknowledgements

We are grateful to Mr J. D. Hamilton and to the Clyde River Purification Board for permission to use their unpublished data on Loch Lomond. Manuscript drafts were typed by Mrs M. S. Wilson and Mrs D. McIntyre.

References

Bailey-Watts, A. E. & Duncan, P., 1981. The ecology of Scotland's largest lochs: Lomond, Awe, Ness, Morar and Shiel. Ed. P. S. Maitland. Chemical characterisation, a one year comparative study. Chap. 3. This volume.

Buchan, A., 1872. Remarks on the deep-water temperature of Lochs Lomond, Katrine and Tay. Proc. R. Soc. Edinb., 1871–72, 791–795.

Buchanan, J. Y. 1885. On the distribution of temperature in Loch Lomond during the autumn of 1885. Proc. R. Soc. Edinb., 1885–1886, 13: 403–427.

Christison, R., 1871. Observations on the fresh-waters of Scotland. Proc. R. Soc. Edinb., 7: 567.

Glover, R. S., 1953. The Hardy plankton indicator and sampler: a description of the various models in use. Bull. Mar. Ecol. 4: 7–20.

Gorham, E., 1958. The physical limnology of Northern Britain: an epitome of the bathymetrical survey of the Scottish freshwater lochs, 1897–1909. Limnol. Oceanogr. 3: 40–50.

Hutchinson, G. E., 1957. A treatise on limnology. New York: Wiley.

Macan, T. T., 1970. Biological studies of the English lakes. London: Longman.

Macan, T. T., 1977. A key to the British fresh- and brackish-water gastropods. Freshwater Biological Association, Scientific Publication No. 13.

Maitland, P. S., 1972a. A key to the freshwater fishes of the British Isles with notes on their distribution and ecology. Freshwater Biological Association. Scientific Publication No. 27.

Maitland, P. S., 1972b. Loch Lomond: man's effects on the salmonid community. J. Fish. Res. Bd. Can. 29: 849–860.

Maitland, P. S., 1978. Biology of fresh waters. Glasgow: Blackie.

Maitland, P. S., 1980. Scarring of whitefish (Coregonus lavaretus) by lampreys (Lampetra fluviatilis) in Loch Lomond. Can. J. Fish. Aquat. Sci. 37: 1981–1988.

Maitland, P. S., 1981. The ecology of Scotland's largest lochs: Lomond, Awe, Ness, Morar and Shiel. Ed. P. S. Maitland. Introduction and catchment analysis. Chap. 1. This volume.

Maitland, P. S., Smith, B. D. & Dennis, G. M., 1981. The ecology of Scotland's largest lochs: Lomond, Awe, Ness, Morar and Shiel. Ed. P. S. Maitland. The crustacean zooplankton. Chap. 6. This volume.

Maulood, B. K. & Boney, A. D., 1980. A seasonal and ecological study of the phytoplankton of Loch Lomond. Hydrobiologia 71: 239–259.

Morrison, J. T., 1886. On the distribution of temperature in Loch Lomond and Loch Katrine during the past winter and spring. Rep. Brit. Ass., 1886, 528.

Mortimer, C. H., 1955. Some effects of the earth's rotation on water movement in stratified lakes. Verh. int. Verein. theor. angew. Limnol. 12: 66–77.

Murray, J., 1887. Depth of Loch Morar. Scot. Geogr. Mag. (Edinb.) 3: 316.

Murray, J. & Pullar, L., 1910 Bathymetrical survey of the freshwater lochs of Scotland. Edinburgh: Challenger. Vols. 1–6.

Pearsall, W. H., 1921. The development of vegetation in the English lakes, considered in relation to the general evolution of glacial lakes and rock basins. Proc. R. Soc. Lond., B, 92: 259–284.

Ratcliffe, D., 1977. A nature conservation review. Cambridge: University Press. Vols 1 & 2.

Rigler, F. H., 1972. The Char Lake project. A study of energy flow in a high arctic lake. Proc. IBP-UNESCO Symp., Kazimierz Dolny, Poland, 1: 287–300.

Rawson, D. S., 1955. Morphometry as a dominant factor in the productivity of large lakes. Verh. int. Verein. theor. angew. Limnol. 12: 164–175.

Reynoldson, T. B., Smith, B. D. & Maitland, P. S., 1981. A species of North American triclad new to Britain found in Loch Ness. Scotland. J. Zool. 193: 531–539.

Robertson, D., 1888. Observed depths in Loch Lomond. Trans. Nat. Hist. Soc. Glasg, 2: 141–143.

Russell-Hunter, W. D., 1958. Biology in the Clyde and its associated waters. In: Miller, R. & Tivy, J. 1958. The Glasgow region. Edinburgh: Constable. 97–118.

Russell-Hunter, W. D., 1970. Aquatic productivity. New York: Macmillan.

Sakamoto, M., 1966. The chlorophyll amount in the euphotic zone in some Japanese lakes and its significance in the photosynthetic production of phytoplankton communities. Bot. Mag. Tokyo. 79: 77–88.

Slack. H. D., 1957. Studies on Loch Lomond, 1. Glasgow: Blackie.

Slack, H. D., 1965. The profundal fauna of Loch Lomond, Scotland. Proc. R. Soc. Edinb., B. 69: 272–297.

Smith, B. D., Maitland, P. S., Young, M. R. & Carr, M. J., 1981a. The ecology of Scotland's largest lochs: Lomond, Awe, Ness, Morar and Shiel. Ed. P. S. Maitland. The littoral zoo-benthos. Chap. 7. This volume.

Smith, I. R. & Lyle, A. A., 1979. The distribution of fresh waters in great Britain. Cambridge: Institute of Terrestrial Ecology.

Smith, I. R., Lyle, A. A. & Rosie, A. J., 1981b The ecology of Scotland's largest lochs: Lomond, Awe, Ness, Morar and Shiel. Ed. P. S. Maitland. Comparative physical limnology. Chap. 2. This volume.

Spence, D. H. N., 1967. Factors controlling the distribution of freshwater macrophytes with particular reference to the lochs of Scotland. J. Ecol. 55: 147–170.

Thienemann, A., 1925. Die Binnengewässer Mitteleuropas. Stuttgart.

Thorpe, S. A., 1971. Asymmetry of the internal seiche in Loch Ness. Nature, Lond. 231: 306–8.

Thorpe, S. A., 1974. Near-resonant forcing in a shallow two-layer fluid: a model for the internal surge in Loch Ness. J. Fluid Mech., 63: 509–527.

282

Thorpe, S. A. & Hall, A. J., 1974. Evidence of Kelvin-Helmholtz billows in Loch Ness. Limnol. Oceanogr. 19: 973–976.

Thorpe, S. A., Hall, A. & Crofts, I., 1972. The internal surges in Loch Ness. Nature, Lond. 237: 96–98.

Thorpe, S. A., Hall, A. J., Taylor, C. & Allen, J., 1977. Billows in Loch Ness. Deep Sea Research 24: 371–379.

Vollenweider, R. A., 1968. Scientific fundamentals of the eutrophication of lake and flowing waters, with particular reference to nitrogen and phosphorus as factors in eutrophication. Tech. Report, Water Mgt. Res. OECD.

Vollenweider, R. A., 1977. Advances in defining critical loading levels for phosphorus in lake eutrophication. Mem. Ist. ital. Idrobiol. 33: 53–83.

Watson, E. R., 1903. Internal oscillation in the waters of Loch Ness. Nature, Lond. 49: 174.

Watson, E. R., 1904a. Movements of the waters of Loch Ness, as indicated by temperature observations. Geogr. J. (Lond.) 24: 430.

Watson, E. R., 1904b. On the ionization of air in vessels immersed in deep water. (L. Ness). Geogr. J. (Lond.) 24: 437–440.

Wedderburn, E. M., 1904. Seiches observed in Loch Ness. Proc. R. Soc. Edinb., 25: 25.

Wedderburn, E. M., 1907a. The temperature of the freshwater lochs of Scotland, with special reference to Loch Ness. Trans. R. Soc. Edinb., 45: 407.

Wedderburn, E. M., 1907b. The temperature of the freshwater lochs of Scotland, with special reference to Loch Ness. Proc. R. Soc. Edinb., 27: 14–15.

Wedderburn, E. M. & Watson, W., 1909. Observations with a current meter in Loch Ness. Proc. R. Soc. Edinb., 29: 619–647.

Index*

---

barrier *264.*
base flow 36.
base-poor *265.*
base-rich *265, 267.*
base rich rock 73.
base richness *136.*
basins 13.
bathing 281.
bathymetric surveys 136.
bathymetry 92, 280.
bedrock 85, *160, 161, 162, 163, 164, 171.*
beetles 181.
Ben Lomond **277.**
benthic invertebrate 159.
benthos *160, 161, 162, 163, 164, 257.*
Bezzia 213, 214.
bivalve molluscs 211, 214, 216, 218.
Bivalvia 158, *172.*
Blelham Tarn 82, 112, 113.
blue-green algae 93, 98, 101, **102,** 107.
blue light *82.*
boat hire 281.
body length 187, *188, 189, 190, 191, 192.*
Borland (fish lock) 232, 233.
Bosmina coregoni 138, *140, 141, 142,* **143,** 145, *146, 150,* 269.
Botryococcus 103.
Botryococcus braunii 101.
boulders **159,** *160, 161, 162, 163, 164, 171,* 178.
bracken **168.**
brackish 228.
Bracora **123.**
breadth *6, 32, 73, 121.*
breaker line 55, *59.*
bridges *22.*
British Isles 70, 228.
British Waterways Board 2, 40, 64.
brook lamprey *234,* 237, *245.*
bryophytes 121.
Building Research Station
Bumilleriopsis 101, 105.
Bythotrephes longimanus 138, *140, 141, 142, 146, 147, 150,* 269.

Ca *74, 77, 78, 79, 107, 156, 257.*
cabin cruising 281.

CaCO₃ *268.*
Caenis **197.**
Caenis horaria *173, 176, 177, 179,* 180, 196, 198.
Caenis moesta *173, 176, 177, 179,* 180, 198, *200.*
Cairngorms 70.
calciphiles 196.
calcium *76,* 79, 198, 209, *210,* 211, 217, *268.*
Caledonian *14, 17, 35.*
Caledonian Canal 13, **38,** 156, **274.**
Callicorixa praeusta 174, 176.
Callitriche hermiphroditica 131.
Callop *14, 19, 35.*
Callop River 227.
Caltha palustris 129.
camping 281.
Canada *262,* 263.
Canadian Shield 70.
canoeing 281.
Capnia atra 174, 177, 179, 182.
Capnia bifrons 173, 176, 177, 179, 182, 198, *200.*
carbon *74,* 81, *156.*
Carex acutiflorus 129.
Carex nigra 129.
Carex rostrata 131.
Carex vesicaria 128, 131.
Carlo Erba Carbon Analyser *74.*
Carum verticillatum 129.
catchment 1, **3,** 4, *6, 14,* 18, *21,* 24, 36, *73, 136, 156, 224, 225, 248, 257, 267.*
catchment area *136, 156.*
cations *74, 78,* 80, *107.*
Ceannacroc 223.
Centrales *99.*
Centroptilum luteolum 173, 175, 177, 179, 180, 182, 184, 185, 196, 198, 199, *212.*
ceratopogonid 214.
Ceratopogonidae 213.
Chaoboridae *212,* 218.
Chaoborus 211.
Chaoborus flavicans 138, *212,* 214, **215.**
Char Lake *262,* 263.
characterisation *140, 200.*
charr 151, *234,* **235,** *240, 245,* 247, 248.
chemical factors 107.

288

289

shore zone   34, *43*, 55, *59*, *63*.
Shropshire   113.
Si   *268*.
Sialis lutaria   *174, 176*, 181.
Sida crystalina   *173, 175*.
Sigara   270.
Sigara distincta   *174, 176, 177, 180*.
Sigara dosalis   *174, 176, 177*, 11, *200*.
Sigara falleni   *174, 176*.
Sigara fossarum   *174, 176*.
Sigara scotti   *174, 176, 177, 200*.
silica   108, 112.
silicate-silicon   80.
silicon   75, 81, *82, 84*, 209, 210, 211, 218.
silt   **159,** *160, 161, 162, 163, 164, 171, 175,*
   178, 209, *210*.
Simulium   138, 196.
SiO₃-Si   *75*.
Siphlonurus lacustris   *175*, 180, *182*, 199, *200*.
   size   254.
slope   4, *21, 34*, 43, *224*.
Sloy Power Station   *226*.
Snaid Burn   *226, 265*.
snail   196, 262.
snow   31.
sodium   *76, 79*.
soft water   196.
solar radiation   31, 45, *48*, 52.
solution   *264*.
Sparganium minimum   131.
spats   185.
species associations   *245*.
species diversity   172, 174, 183.
spectral quality   74.
Sphaerocystis   **102.**
Spondylosium   **102.**
sport fishery   241.
Stampellina   213.
Staurastrum   *100,* **102,** 103.
Staurastrum cuspidatum   100.
Staurastrum gracile   100.
Stephanodiscus astraea   98.
Stictochironomus   213.
stoneflies   180, 181, 196, 199.
stone loach   *234*, 237, *245*, 260.
stone-kicking
stones   **159,** *160, 161, 162, 163, 164, 171,* 178,
   179.

stone-washing
storage   37.
storm   **56.**
storm beach   **56, 57.**
stratification   49, *63*, 102, 258.
stream junctions   4, *21*.
streamflow   35.
streams   198, 199, 260.
Stylaria lacustris   *173, 175, 177, 179, 181*, 196.
Stylodrilus   270.
Stylodrilus heringianus   *173, 175, 177, 179,*
   181, 196, 211, *212*, 214, 216, 220.
sub-catchment   4, *14*, 15, *20*, 24, *35*.
submerged   121.
subsamples   93.
substrate   158, **159,** *160, 161, 162, 163, 164,*
   170, 178, 183, 196, 248, 259, 260.
Subularia aquatica   128, 130, 131.
sunshine   30, 47, *48*.
surface area   *136, 224*.
surface waters   77.
survey   279.
suspended solids   75, *268*.
swash zone   *59*.
Sweden   229.
synthesis   253.

Tabellaria   93, *99*, **102,** 111, 112.
Tabellaria fenestrata   98, 103, *104, 106*, 107.
Tabellaria flocculosa   98, 103, *106*, 111.
Tanganyika   *261*.
Tanypodinae   218.
Tanytarsus   211, *213*, 214, 216, 220, 270.
Tanytarsus signatus   211.
Tarbet   **122,** *127*, 130.
Tasmania   70.
tectonic   *264*.
temperate   *264*.
temperature   30, 45, 47, *63*, 102, 218.
temporal variation   77.
tench   243.
ten-spined stickleback   *234*, 237, *245*, **246.**
thermal   *264*.
thermal characteristics   47, *50*.
thermal stratification   103.
thermocline   *63, 92*, 93, *267*.
Theromyzon tessulatum   *175*.
three-spined stickleback   151, *234*, 237, *245*,
   246.